The Dutch National Research Agenda in Perspective

The Dutch National Research Agenda in Perspective

A Reflection on Research and Science Policy in Practice

Edited by
Beatrice de Graaf, Alexander Rinnooy Kan
and Henk Molenaar

Amsterdam University Press

Cover illustration: DimarDesign

Cover design: Coördesign, Leiden
Lay-out: Crius Group, Hulshout

Amsterdam University Press English-language titles are distributed in the US and Canada by the University of Chicago Press.

ISBN   978 94 6298 279 6
e-ISBN   978 90 4853 282 7 (pdf)
DOI   10.5117/9789462982796
NUR   600

Creative Commons License CC BY NC ND (http://creativecommons.org/licenses/by-nc-nd/3.0)

☺ All authors / Amsterdam University Press B.V., Amsterdam 2017

Some rights reserved. Without limiting the rights under copyright reserved above, any part of this book may be reproduced, stored in or introduced into a retrieval system, or transmitted, in any form or by any means (electronic, mechanical, photocopying, recording or otherwise).

Contents

Foreword by Jet Bussemaker, Minister of Education, Culture and Science   9

Introduction   11
   Beatrice de Graaf, Henk Molenaar, and Alexander Rinnooy Kan

The Art and Science of Asking Questions   19
   José van Dijck

A Plurality of Voices   31
   The Dutch National Research Agenda in Dispute
   Henk Molenaar

National Research Agendas   47
   An International Comparison
   Wim de Haas

The Role of Universities of Applied Sciences in Implementing the Dutch National Research Agenda   61
   Daan Andriessen and Marieke Schuurmans

Steering Scientific Research and Reaping its Benefits   75
   Reflections on Dutch Science Policy
   Coenraad Krijger and Maarten Prak

Managing what Cannot be Managed   87
   On the Possibility of Science Policy
   Barend van der Meulen

The Art of Making Connections   101
   Ed Brinksma

Too Big to Innovate?   121
   The Sense and Nonsense of Big Programmatic Research
   Brian Burgoon, Marieke de Goede, Marlies Glasius, and Eric Schliesser

The Art of Asking Questions, and why Scientists Are Better at it 137
    Herman van de Werfhorst

Skip the Agenda Building 147
    Let the Wisdom of the Crowd Drive a Dynamic Tapestry of Science
    Marten Scheffer and Johan Bollen

An Economic Perspective on the Dutch National Research Agenda 155
    Roel van Elk and Bas ter Weel

What is the Good of Government Interference in Science? 167
    A Question from Late Nineteenth-Century Germany
    Herman Paul

Free-range Poultry Holdings 181
    Living the Academic Life in a Context of Normative Uncertainty
    Beatrice de Graaf

A National Research Agenda and the Self-understanding of Modern Universities 193
    Rutger Claassen and Marcus Düwell

No University without Diversity 209
    The Dynamic Ecosystem of Scientific and Social Innovation
    André Knottnerus

Inspiration 221
    Louise Gunning-Schepers

Process of Developing the Dutch National Research Agenda 225

Index 239

List of Tables and Figures

Tables

Table 1	Characterisation of innovation policy in several countries	49
Table 2	National research prioritization: characterization for fifteen countries	51
Table 3	Prioritized themes in national research agendas (in italics: themes mentioned five times or more; bold: some notable research themes, for various reasons)	53
Table 4	Process of development and implementation of national research agendas	55
Table 1	Key figures for Dutch universities of applied science (2014)	62

Figures

Figure 1	The linear model	103
Figure 2	Stokes' quadrants	105
Figure 3	Stokes' dynamic model	107
Figure 4	Extension of the Stokes model	109
Figure 5	Squaring the golden triangle	116

Foreword by Jet Bussemaker, Minister of Education, Culture and Science

As I see it, imagination and connection are the two most important characteristics of science and the Dutch National Research Agenda.

Imagination is a vital prerequisite for developing new perspectives in scientific research. After all, we are now developing the knowledge we will need in the future.

Imagination finds expression in the vast number of questions posed by scientists, citizens, businesses, and civil society organisations as part of the Dutch National Research Agenda – close to 12,000 questions on a wide variety of themes and topics. These range from the human significance of art to organising healthcare on the basis of unique and individual characteristics, showing the depth of people's interest in and commitment to the world of science: one of the main reasons why the Dutch National Research Agenda exists.

The questions the Agenda addresses also illustrate the challenges that science is facing in the years to come, and the essential nature of connection and cooperation between sectors and disciplines.

If we want to find answers that have an impact on society, I firmly believe that we need to come up with new, creative combinations. Combinations between technology and art, between historical roots and futuristic concepts, facts and imagination, science and the working world, new and existing knowledge. The Dutch National Research Agenda invites us to make these connections and to embrace cooperation throughout the chain of knowledge, in particular with society.

One inspiring example is the collaboration between museums, universities, universities of applied sciences, and industry, aimed at developing innovative products on the basis of long-established museum collections. One research group is using the collection of the Naturalis Biodiversity Centre in Leiden to develop natural sweeteners, which can be sustainably cultivated and help prevent diabetes. Museum collections are also a source of expertise, for example helping customs officials to detect endangered wood species in musical instruments.

The essays in this volume, all of which reflect upon the Dutch National Research Agenda, can be seen as an ode to imagination and connection. They are also an ode to critical inquiry, encouraging us to continue to interrogate the types of questions we ask.

Dr Jet Bussemaker
Minister of Education, Culture and Science of the Netherlands

Introduction

Beatrice de Graaf, Henk Molenaar, and Alexander Rinnooy Kan

Asking questions

'What is the proper use of the word "no" and what isn't?' 'Would it be possible to create a funicular to the moon?' Questions like these are more likely to be asked by curious students or children than by sophisticated researchers. And yet this type of unbounded curiosity remains one of the main drivers behind fundamental scientific research. That is why these and nearly 12,000 other questions were all admitted onto a nationwide platform with the intent to aggregate the national curiosity of the Dutch – a platform that was designated to become the Dutch National Research Agenda.

Both the agenda's format and process were unique in their kind. All earlier national efforts undertaken in other countries had opted for a top-down format, in which the customary committee of wise advisors produced a respectable but rather predictable outcome. The bottom-up approach favoured in the Netherlands was hotly contested and heavily debated. But in the end, it produced a rich research menu, identifying a range of issues that appeal to the research community as well as to the general public (see Annex for a description of the process of developing the Dutch National Research Agenda).

Thus, one of the characterizing features of the Dutch National Research Agenda was precisely that it was created through public consultation. Nowadays, this sort of consultation is used commonly in a variety of areas. It is, of course, used by business enterprises to assess and gauge consumer preferences and desires, and it also figures in political decision-making processes such as crafting a national referendum, or in other forms of participatory democracy. As such, the format is not new at all. However, for academic science and research, 'citizen science' is a relatively new notion. Crowdsourcing has only recently become a resource for long-term funding for new research. As Ed Brinksma points out in his contribution, the use of the internet has irrevocably speeded up and expanded public engagement with academic research and innovation far and wide. Increasingly, research projects do not only take shape through the interaction of government, science, and industry; citizens – be they amateur scientist, investor, consumer, societal stakeholder, inventor, or entrepreneur – and the public at large have become contributing voices as well.

The desire to provide public knowledge, to generate scientific insights for and with society and industry, is not a new phenomenon as such. Throughout the centuries, the university was the very place where clerical elites, politicians, state representatives, and diplomats were educated, preparing them to assume their role in the power system of the day. When the Dutch universities were liberated from French rule and restored to national autonomy in 1815, a system of higher learning and academic research was established that was geared towards 'producing a learned elite for the country'. It was, at the time, most notably staffed with theologians, philosophes, *hommes des lettres*, and a few medical professors, mathematicians, and physicists. Research back then responded to demands from the public domain, in particular from the newly created seats of power and administration, but also from the churches. Large chunks of the government's research budget were allocated to the salaries of theological professors (two thirds of Utrecht University's students were theologians, aspiring to the clerical robe). In 200 years, academia has shifted gears. Today's science policymakers respond much more to requests from industry and commerce. They tend to stress the importance of 'Science Parks' for research and innovation in the natural and life sciences.

Not everything from the past needs to be preserved, nor does every recent research innovation call for emulation. It is also undoubtedly the case that research projects today are being influenced by a widely expanded audience, and that researchers themselves are confronted with many more conflicting demands than they have ever been before. Since 1945, society's role and the citizen's place with respect to institutions of higher education and academic research have grown: the general public is eventually the ultimate recipient of scientific findings; parents send their sons and daughters off to university; a sizeable portion of citizens' taxes helps fund the national research and teaching budgets. Not surprisingly, the populace demands something in return. But what exactly?

Since the Enlightenment, modern universities and research institutes have undergone a Baconian revolution, placing professionalization of academic standards, disciplinary differentiation, and specialization at the zenith of their ambitions. Only when science is first and foremost allowed to render service to science itself and formulate its own questions, the conventional wisdom says, will it be able to open new horizons and optimally serve society and industry in its wake. Science does not simply respond to already formulated questions, it invents and formulates new ones, answers needs and concerns that were not there before. Today's graduate and postgraduate students are therefore trained simultaneously to work towards professionalization and specialization, on the one hand, and to transcend boundaries and share their insights with society, on the other.

The art of asking the right questions is therefore exactly that: an art, combining hunches and sound professional disciplinary knowledge with a long-term dedication to unleash creative energy to meet the needs and concerns arising from the public or from commerce and industry. In this context, the emergence of a new kind of 'citizen science' – of new instruments to involve and mobilize the public – does not come as a surprise. Today's societies are highly educated and perfectly able to act not just as benefactors of science, but as co-creators of research needs, aims, and constraints as well.

Academic research has, to a notable degree, always been a public service. But in opening up the Dutch National Research Agenda to the public, the public voice in the bottom-up articulation of programming science has been made more explicit and visible as a channel of influence in its own right. In this volume we will further explore, debate, and contest the arrangement between science, industry, government, and the public in generating research.

Asking questions – *sapere aude!* – is one of the core ingredients of becoming an adult, of transcending existing cognitive constraints. In that spirit, questions are also being asked in this volume regarding the uses, benefits, challenges, and risks of creating and having a research agenda, about the scope of research policy itself, and concerning the ways in which government involvement in research and scholarship can and should work – or not.

Structure of this volume

In this volume the making of the Dutch National Research Agenda is described as a case study of a new way of asking questions and of combining research and the public domain, but it is also intended to critically evaluate the desirability and (im)possibility of steering science as such. Can/may the public intervene from the outside in the inner world of research dynamics? Is allocating budgets a one-way street? Should science decide on its own, citing the so-called Haldane principle, on how to spend these precious public resources? The process of crafting the platform for the Dutch National Research Agenda inspired various rounds of debates, criticisms, and reflections on the use and nature of science and on the entanglement of science, science policy, and the public, thereby contributing to a lively atmosphere of academic discussion. This volume is an attempt to unravel these discussions and make them accessible to a larger public of interested citizens, scientists, and policymakers in the Netherlands and abroad.

This volume is structured around three strands in the debate that surfaced between 2014 and 2016, while the agenda was being created: 1) the process of developing the agenda as such, 2) the (im)possibility of steering science, and 3) the use of science in a wider philosophical and historical context.

The first part of this volume is dedicated to the process of agenda-setting. José van Dijck, President of the Royal Netherlands Academy of Arts and Sciences (KNAW), takes the lead in highlighting how the agenda became a national exercise in asking science 'researchable' questions. For her, asking the 'right' research questions is one of the highest arts in academia. She explains how the agenda offered a platform that triggered 'new collective insights, unexpected alliances, and novel routes through known territories'.

Henk Molenaar, secretary to the Dutch National Research Agenda, describes how the agenda was launched, and how it set out to establish 'big questions' and forge interrelationships between the multifarious research programmes of universities, research institutes, private sector companies, and other knowledge organisations. He identifies three nodes of debate that permeated the whole of the agenda-setting process: the relation of the agenda to unfettered research, the tension between disciplinary diversity and thematic focus, and the question of legitimacy and public support. Is science inherently legitimized in open, democratic societies or does it benefit from explicit public involvement?

This agenda-setting process is put into a wider, international context by Wim de Haas of the secretariat of the Dutch National Research Agenda, who examines practices of thematic research prioritization in various countries. Daan Andriessen and Marieke Schuurmans focus on the place of the universities for applied science, or colleges (*hogescholen*) within this process, institutions of higher learning that sometimes tend to be overlooked in scientific research debates. According to them, these colleges are very well-positioned to participate in the task of focusing and clustering: '[Their focus] on practice-oriented research and their strong network in professional practice will ensure that the National Research Agenda truly contributes to society'.

In the second part of this volume, the (im)possibilities of intervening with and steering science are debated. The chapters here echo the intense academic and public debate during the process of the agenda-setting activities. Maarten Prak and Coenraad Krijger, from the perspective of the Netherlands Organisation for Scientific Research (NWO), underscore the

INTRODUCTION 15

fundamental problem of science policy: the fact that 'results of research projects cannot be predicted (because if they were, research would be futile)'. So, how – given the prospect of unpredictable results – can huge sums of public money be spent legitimately and wisely? Their contribution presents an illuminating overview of types of science policy and various dimensions of research impact. In addition, Barend van der Meulen (Rathenau Institute) further elaborates on this theme by comparing science policy with a principal-agent game, in which all players have to cooperate in order to minimize uncertain outcomes as well as the risks of wasting scarce resources.

Next, the Rector of the University of Twente in Enschede, Ed Brinksma, highlights the importance of making connections. For universities of technology, the research portfolio is of course heavily influenced by application domains and stakeholders in industry and society. Brinksma offers a model for approaching the connections between different types of research and science policy. He points out that 'successful research policy is an art of making the right connections: connections between Bohr, Pasteur and Edison types of research, between research and education, with the agendas of regional, national, and supranational governments, with the priorities of industry, and, increasingly, with the preferences of the public'. To boost research and innovation, investments are needed in all of the disciplines – from technology to the humanities, from applied to blue skies research – and most of all in furthering the connections between these different types of research.

From a wholly different angle, Brian Burgoon, Marieke de Goede, Marlies Glasius, and Eric Schliesser, all professors of Political Science at the University of Amsterdam and recipients of large grants from the NWO or the European Research Council (ERC), argue that the tendency of awarding ever larger grants undermines the dynamics of research diversity. Large grants to ever tinier shares of submitted research proposals impose a rat race of winners and losers onto the community of researchers and demoralise promising young scholars. Science policy should therefore also determine a broadening of the available grant mix, as well as a diversification of societal stakeholders participating in the process of agenda-setting.

Bas ter Weel, from the Netherlands Bureau for Economic Policy Analysis, brings an economic perspective to the table and approaches the issue of research steering from the angle of market failure. He ponders the balance between the risk of scattered research and underutilization of complementarities on the one hand, and the far too conservative or market-driven economies of scale on the other. Marten Scheffer and Herman van de Werfhorst round off this session with provocative pleas for the total abandonment of top-down science planning (Scheffer) and for an equal division of the research budget among individual researchers for them to redistribute amongst themselves and their colleagues (Van de Werfhorst). This revolutionary plan should be read in conjunction with the latter's scepticism vis-a-vis the alleged wisdom of society's competency to allocate resources as compared to that of the scientists themselves.

The third section of this volume zooms out to embrace a wider vista on the question of good governance in science. What is the aim or purpose of the university and of research? Historian Herman Paul makes a case for the reintroduction of the language of vice and virtue in the debate on 'aims of science'. Rather than to profitable outcomes, academic self-management, or an equal division of resources, attention needs to be given to the attitudes, ethics, and habits of researchers and scientists. Good science needs to be historicized and the aims of science have to be put in perspective. Only then will we be able to acknowledge that questions about the aims of science are inherently moral ones.

Paul's argument for opening up the debate to moral questions is further elaborated upon by Beatrice de Graaf's (historian and terrorism expert) analysis of the normative uncertainty underlying the debates and disputes on science policy and legitimacy mentioned above. She outlines two narratives that seek to clarify the academic life and its purpose: the utilitarian 'goose model' (or 'goose with the golden eggs') and the Humboldtian '*Bildung* model'. She shows how the ideas, goals, and expectations of each model continue to compete for recognition and endorsement. And although the former is currently gaining the upper hand, the values of the other model are essential to sustaining the life of the mind. This conflict of values regarding science and the scientist is precipitating a significant degree of uncertainty in politics, academia, and society regarding the aspirations of the academic endeavour. De Graaf makes a case for restoring the balance by acknowledging and defending the diversity and richness of the academic lives at stake, and by countering moves that may cause one vision to monopolize all others.

Philosophers Marcus Düwell and Rutger Claassen continue this line of thinking. While arguing that scientific research is fundamentally about the self-understanding of human beings, they confirm that communal forms of priority setting are sought after since the task of interpreting ourselves is a collective, not an individual one. However, they question the democratic character of the current exchange between scientists, politicians, and policymakers on the one hand, and a wider group of private (especially corporate) interests on the other, and call for a 'new relationship between the roles of political institutions, societal interest groups, and the researchers themselves'.

Before Louise Gunning, chair of the Dutch National Research Agenda since 2016, closes this volume with an epilogue, André Knottnerus, President of the Netherlands Scientific Council for Government Policy (WRR), pays tribute to the fine-grained, delicate 'ecosystem' of the Dutch research environment and advocates better protection and more respect for this system of diversity.

An open invitation to connect

To sum up and invite the reader to ponder the preceding arguments, the chapters might be summarised as a collective attempt to highlight the importance of stimulating national and international curiosity, and doing so in a well-balanced, legitimate, democratic, and reflective manner. If we want science and society to move forward and to remain in flux, this infinite curiosity has to be propelled by inquisitive minds finding each other, working together, and transcending boundaries. At the end of the day, the inventory of national curiosity that the agenda set out to be miraculously transformed itself into a treasure trove of broad, mostly multidisciplinary and multi-sector research questions that derive additional legitimacy from the bottom-up way in which the agenda was construed. In a research environment as sophisticated and well-positioned as in the Netherlands, possibly the greatest potential to be unlocked lies in finding a new balance between deep scientific specialization and broad societal interests. The Dutch National Research Agenda might well serve to illustrate these opportunities to a European or global audience in need of a similar innovation.

The Art and Science of Asking Questions

José van Dijck

'The art and science of asking questions is the source of all knowledge.'
(Thomas Berger)

Last year, I received a rather desperate email from a 16-year-old secondary school student, who was wondering if I could help with her project thesis. Could she please interview me about the power of media? She wanted to know whether government censorship of mass media could contribute to curbing the threat of terrorism. Newspapers and television news, she assumed, were instrumental in spreading ideas of violence. If we could only find out how the media steer public opinion, we could do something about that threat.

It is not uncommon for secondary school students – or undergraduate students for that matter – to start an academic project with a wide-ranging question when they only have a vague sense of what they are looking for. Each time I receive a request like the one above, I realise how difficult it is for young students and aspiring scholars to articulate the 'right' question: right in terms of scope, ambition, and context. Time and again, it turns out to be very difficult to find a thesis that is not only interesting and challenging, but also practical and doable within a set timeframe. A broad question often indicates the boundless inquisitiveness of a young student's mindset, but it just as much connotes his or her helplessness in shaping the immense world of potential topics. Curiosity gallops away, untamed by the pragmatism of academic scholarship, which only sets in after years of professional training.

As every academic knows, no thesis can be successfully pursued without specific limitations, no dissertation can be written without first setting the terms of reference: a precise research question that limits the scope of the subject matter to be covered, and that allows one to select a methodology and to tackle the practicalities of the execution. Good scientists know how to tackle 'Big Questions' by breaking them up into smaller ones, each addressing a partial and manageable problem, setting realistic goals within the confines of time and space. Excellent scientists do the same, but they also know how to translate smaller questions back to the overall Big Question, adding a major insight to a daunting problem. Big Questions require practice in small thinking without losing sight of the big picture as well as a sense of urgency.

My problem with the above-mentioned student's question was not so much its content, but its scope in relation to the context of a secondary school project. So I returned her email, suggesting she make her research question 'smaller', e.g. by relating it to one or two concrete examples. I advised her to compare several newspaper and TV reports covering a terrorist attack and analyse the verbal and visual rhetoric used to describe the motives and backgrounds of perpetrators. By selecting a specific case, and focusing on specific rhetorical expressions, she might be able to get a handle on the larger problem of mass media affecting public opinion. How much time did she have available to do the research and write her paper? And by the way, by making her question more specific I did not mean to discourage her from asking Big Questions; the last thing I wanted was to stymie her curiosity.

The idea that asking questions is one of the highest arts in academia usually triggers a sceptical response: is not the ultimate goal of science to come up with answers? But science is not prophecy: what distinguishes scientists from oracles is their ability to raise the right question before they start researching an appropriate answer. Perhaps more accurately: articulating the right question at the right time in the right context is a sine qua non for successful research. It takes time and effort to teach students to 'tame' their curiosity into a manageable process; articulating a 'researchable' question requires finding a balance between inquisitiveness and practical constraints. Or, as the Irish poet James Stephens beautifully phrased it: 'We get wise by asking questions, and even if these are not answered, we get wise, for a well-packed question carries its answer on its back as a snail carries its shell' (Stephens, 1920, p. 37).

The Dutch National Research Agenda: starting from questions

The making of the Dutch National Research Agenda (*Nationale Wetenschapsagenda*, or NWA) turned out to be precisely such a balancing act between the potentiality of endless questions and the reality of relentless constraints. In the spring of 2015, Dutch citizens were asked to send in questions worthwhile for researchers to tackle, examine, test, or solve. There were few restrictions as to what kind of questions people could ask: they had to challenge researchers, were preferably original and new – in the sense of 'unanswered' – and should be 'researchable' over the next ten years by research groups funded in the Netherlands. A mere technical condition was that questions had to be written in Dutch, simply because this was a national agenda. Such were the conditions for enabling a bottom-up process of agenda-setting. A national research agenda that stemmed from questions raised by citizens, rather than themes imposed by governments or industries, was a truly novel idea. The Agenda became a national exercise in asking 'researchable' questions to science. In less than a month, almost 12,000 questions were collected through a special website (Nationale Wetenschapsagenda, 2015). But what makes a question researchable?

There is no formula or recipe for what a good researchable question is because, first, that definition depends on the field or discipline and, second, asking questions is not a *thing* but a process, a skill, and a growing insight all at the same time – indeed, a package to be carried on the snail's back. What happened to these 12,000 questions collected in April 2015 involved a process of *filtering, categorizing, 'packaging', reassembling*, and *prioritizing*. That process was professionally managed and supervised by a large group of diverse and qualified researchers, and coordinated by the Royal Netherlands Academy of Arts and Sciences. Almost seventy researchers selected from a range of disciplinary backgrounds, age, and gender collaborated in five teams, roughly representing the social sciences, humanities, technical sciences, natural sciences, and medical and health sciences. In early May, they set out to filter, categorize, package, reassemble, and prioritize.

Filtering questions

The first step, filtering for invalid questions, was the simplest task, even if daunting considering the large number of queries. Surprisingly, citizens submitted remarkably few absolute bogus questions. Some questions did not mean anything; 'What is the secret of the moon?', for instance, sounds more like the title of a children's book than a question to science. Some questions were phrased like riddles, awaiting a prepared answer. And, as expected, some perfectly legitimate questions to science were articulating problems that had already been resolved. Along the same lines, very practical questions that were well underway of being solved, were put aside, for instance: 'Can we develop an MRI-scanner that makes less noise?' is a legitimate technical problem which is currently being tackled by researchers in the medical technology sector.

At the other end of the spectrum, some questions were disqualified because they were too ambitious for the set time frame: 'Can we put men on Mars where they can build a peaceful society?' is a challenge few Dutch scientists would want to accept if the deadline was 2025. What is more, the Mars project would probably be too costly for the Dutch taxpayer, while the Dutch Netherlands lacks an appropriate infrastructure to launch a research programme of this scale. That should not keep us from dreaming, though; without dreams there would be no Large Hadron Collider in Geneva and there would have been no man on the moon in 1969. Big dreams and big questions demand imagination and a sense of urgency. Virtually all pressing problems in the world today are global in scope and require international collaboration to be solved. Many questions raised through the Dutch National Research Agenda addressed global problems requiring Big Thinking. And therefore, they also required 'small' thinking – cutting up large questions into sizeable packages that can be carried on the backs of researchers.

Disqualifying invalid questions is not as easy as it seems: jury members sometimes faced a dilemma whether to put aside a question because it was badly articulated or simply impracticable. For instance, rhetorical questions have that oracle quality many people love to attribute to science. 'Can we realise peace on earth in less than ten years?' 'Why are nations still at war with each other?' or 'Why do people still get sick?' are popular variants of this genre. By the same token, there are a number of so-called million-dollar questions that hold the middle between wishful thinking and shooting on a star. 'Can we transform all polluting carbon dioxides into edible nutrients?' Such questions may be highly challenging and comprise brilliant ideas, and yet they tend to be quite impracticable if you have to work within the set of conditions of time and space. Scientists are not allowed the liberties of science fiction authors. Nevertheless, if it were not for the power of imagination, many inventions would never have found their way into the world.

The task of filtering and weighing the validity of questions turned out to be an important first step: not just to get a sense of people's curiosity and skillfulness in articulating questions, but also to get a better idea of what people think science can do: their expectations, projections, hopes, and resentment. As an academic, it is rather instructive – if not sobering – to find out what powers ordinary citizens attribute to science.

Categorizing and packaging questions

The next step – one that kept the jury members very busy – was to categorize and lump together questions that were similar in nature. This may sound like an easy task, but once the reality of almost 12.000 questions sank in, it became a sizeable challenge. To make the process manageable, questions had to be submitted to one of the five jury teams; when a question was misplaced, it would be turned over to another team. Social Sciences received the most questions (approximately 4,400), followed by Health and Medical Sciences (3,000), Natural Sciences (2,000) Technical Sciences (1,400), and Humanities (1,200) (Van Hintum, 2015). Each of the juries was asked to group questions that showed sizeable overlap and articulate an overarching *question* (not a theme!) for each cluster. Eventually, the five juries combined came up with 140 clusters of grouped questions, each headed by an overarching question that met all conditions for 'researchability'. Or, in the words of the aforementioned Irish poet, the result was 140 'well-packed questions' carried on the backs of scientists.

Most of the 140 packaged questions are illustrative of their overarching umbrella quality. Questions like 'Are religion and modernity each other's opposites?' and 'How can we reduce poverty and increase global well-being of people?' cover a number of questions originally submitted to humanities and social science juries. Other questions, such as 'How does sleep affect health conditions?' cover a large cluster of questions stemming from the health sciences, whereas 'How can we safely store and transport sustainable energy?' clearly emanates from the technical engineering domain. Another of the 140 packages, titled 'Is life possible outside planet Earth?', at first reminds one of the men-on-Mars question; however, it encompasses a number of fundamental questions sprouting from the natural sciences. Finally, a question such as 'Will digitization save our cultural heritage?' formed the umbrella for many inquiries into the effects of digitization on arts and culture. The range of packaged questions spans a large number of disciplines and *types* of science: questions from fundamental as well as applied sciences; 'what', 'how', and 'why' questions; questions that open up broad horizons as well as narrow windows on problems.

Evidently, the process of packaging prompts issues of (trans-)disciplinarity and (cross-)ownership. Indeed, Big Questions and Big Problems are seldom solved by a single discipline or even within a broad academic field such as the humanities or engineering. Science requires specialisms, but it is a fallacy to think that questions 'naturally' fit within a disciplinary scope or belong to a self-evident field of inquiry. On the one hand, the disciplinary jury teams provided a necessary validation framework for weighing and interpreting each question. Recognizing overlap between questions and identifying underlying concerns requires profound knowledge of a scientific field. On the other hand, jury members needed to have a much broader horizon than their own field to judge and weigh questions that defy disciplinary boundaries. For instance, a large number of questions related to the prevention and treatment of a variety of different cancers. To group those questions, jury specialists needed to understand fundamental life science research (e.g. stem cells) as well as technical and chemical facets of cancer treatment. Moreover, including social science perspectives in medical science is often an eye-opener, for example when habits of use or behavioral conditions impact treatment. Tackling a complex problem increasingly requires the concerted effort of scholars from a variety of disciplines.

The categorization process comes at a cost: it takes time and extensive deliberation to discuss the exact articulation of a problem that encompasses a large number of original questions. What happened was a process of intensive give-and-take among the jury members before they finally settled upon the 140 overarching questions. Four months went by before the juries could deliver their 'packaged parcels'. Those four months also included a three-day conference, where almost one thousand academics and interested individuals participated in the process of deliberation. Articulating those packaged questions turned out to be a major challenge in the making of the Dutch National Research Agenda; on the upside, it invited unexpected allies to combine surprising perspectives.

Reassembling Big Questions

Good scientists are quite capable of breaking up big questions into smaller ones; excellent scientists also manage to reassemble the smaller jigsaws to construct a Big Picture. The art and science of asking questions has never been neutral or value free. Asking questions is always also about interests and stakes: for whom is this question important? Who has a stake in raising this issue? Why is this question important now and will it be for the next ten years? Before the NWA process began, three contextual frameworks for asking questions were set by the Ministry of Education: *Science for Science*, *Science for Competitiveness*, and *Science for Society*. Whatever the outcome, taxpayers' money spent on research would have to benefit each of those three areas of interest; and if they were not entirely compatible, they had to be at least complementary.

Many questions submitted to the juries betrayed a sense of urgency: the desire to take on global problems such as climate change, the impact of big data on the organisation of society, or the prevention of terrorism. But almost invariably those big problems were distributed across a fair number of these 140 questions, cutting through disciplines and established research areas.

For example, the question of how to prevent terrorism could be identified in research questions requiring the input of philosophers, historians, economists, social scientists, religious experts, media scholars, computer scientists, and sociologists. Of course all experts bring their own preferences, perspectives, and methodologies to the table; bundled together, these perspectives may offer a concerted effort to tackle a Big Question and come up with a comprehensive approach to a problem.

Identifying so-called routes was the next step in the NWA process; what were the urgent and challenging Big Questions emanating from these 12.000 questions grouped in 140 clusters? The juries set out to identify pressing concerns that cut through all disciplinary jury teams and came up with sixteen exemplary paths. In theory, there are endless potential routes cutting across all fields. By identifying common concerns, these routes showed the way to complementarity and potential collaboration – an invitation for scientists to join communal efforts, to regroup their workforce, and combine their skills to take on a Big Question.

Among the sixteen identified routes are 'Using big data responsibly – searching for patterns in large databases' – an issue that obviously concerns all disciplines and research areas, affecting both society and industry. Other routes are titled 'Personalised medicine' and 'Smart, liveable cities'. Note that each of these routes encompasses an array of questions, often cutting across all research areas, from engineering to humanities, from natural science to law, and from economics to health. Once questions get bigger, their complexity grows in size and the need for cooperation and coordination increases accordingly.

Reassembling packaged questions into routes was another stage in the bottom-up process that the Dutch National Research Agenda turned out to be. Many snails with packs on their backs encountered other snails along the way; sniffing each other's scent, they decided to hook up for a while. Identifying routes was a way for academics to get to know each other by means of pairing questions in order to take on Big Questions. Over the next six months, workshops will be organised to bring together philosophers, engineers working for industrial employers, medical doctors, and lawyers (to name just a few) to settle on collaborative projects and to define their common interests combining diverging perspectives. For some, it is a first-time experience not to be prolonged; for others, it is an opportunity to broaden their professional or academic horizon. For most, the process will be at least an eye-opener to the endless potential and inescapable constraints of 'collaborative science empowered by the people'.

Prioritizing questions

Asking questions to science is more than a beauty contest, where academic pageants are competing for the prettiness of their formulas or the splendor of their arguments. Asking the right questions with the right sense of urgency in the right context is a skill that most successful scientists train to perfection in the course of their careers. Competitive environments have become the natural habitats of academics, certainly in the Netherlands – a country whose track record in securing grants from the European Research Council's schemes is rather impressive (Government of the Netherlands, 2014).

So not surprisingly, many academics initially considered the Dutch National Research Agenda to be a money/power contest – an arena where scientists fight for the visibility of their research fields, filing their claims to fame and their declarations of indispensability. Behind many a question a recognizable world of economic interests and ideological preferences is hiding. If each question submitted to the NWA was a proposition begging for attention and status, some were outright petitions for money and people. Even though the majority of submitted questions came from individual laypersons – including children and students – professional academics submitted a substantial number of them. As stated earlier, asking questions is a trainable skill for students and academics wanting to tame their curiosity. For professional academics, though, submitting research questions is always also a contest in persuasiveness. Most researchers who enter the funding arena have to convince their colleagues as well as many other actors – funding agencies, business investors, lay people, special interest groups, or society at large. Professional researchers are also accustomed to prioritizing each other's proposals: peer review and jury validation is at the core of most science funding schemes.

But how to set priorities in a scheme void of financial rewards and geared toward collaboration rather than competition? In a world where competition is the norm, the kind of wide-ranging collaboration triggered by the NWA was surprising, to say the least. Surprising because no carrots or sticks were put in front of the snails. A vague promise by the Ministry of Education to increase research budgets if the bottom-up process led to better cooperation was never translated into concrete figures. So what drives scientists to collaborate on Big Questions? In the context of a national research agenda, an interesting discussion emerged about priorities. First, does a country *need* to set priorities by selecting a few scientific projects? And second, who decides which research questions are the most interesting, valid, urgent, worthwhile, and practicable for scientists to take on? Is it elected representatives, researchers, juries, self-appointed officials, or citizens who decide what questions scientists take on?

Both questions are in fact highly political and ideological. Most academics will argue that scientific research can and should not be directed by preselected themes that reflect choices made by others – and rightly so. By the same token, many academics are quite eager to contribute their capacities to solve Big Questions; whether this is out of a sense of altruism or out of scientific curiosity is beside the point. While politicians and industries commonly prefer investing money in a few predictable priority areas, the Dutch academic community proposes to prioritize a relatively large number of collaborative projects. The sixteen reassembled 'routes' cover a broad range of topics, leaving room for more. This peculiar choice may confuse many stakeholders, not in the least politicians. If you can't afford to do it all, why don't you come up with a handful of priorities?

Fighting for funding or wrestling for wisdom?

It is at this point that we need to bring up the existential question: what are national research agendas for? Is this effort an exercise in 'citizen science', allowing the people to ask questions to science? Or is it a funding scheme by unusual design? I think one of the most interesting outcomes of this experiment may well be how the NWA is gradually turning into an *instrument* for facilitating bottom-up connections and collaborations. The NWA set in motion a process of first collecting questions, then filtering, categorizing, packaging, reassembling, and prioritizing them. Meanwhile, the *process* is becoming perhaps more valuable than the outcome. Hopefully, the result is a *platform* that is finished – a platform that triggers new collective insights, unexpected alliances, and novel routes through known territories. More than that, the process will exemplify why the power of academia rests not with a handful of (pre)selected disciplines, themes, or brilliant scholars; the power of academia is in its vibrant ecosystem sustained by waxing and waning collectives of researchers from all disciplines and fed by asking questions.

It would be dishonest to conclude that hundreds if not thousands of Dutch researchers who engaged with the NWA one way or another lost interest in the funding game. The next step in the NWA process will be to put forward a claim to the government for a substantial investment in science. This claim will not entail a momentary pecuniary injection into a few trendy topics, but it will call for boosting the ecosystem as such – an ecosystem that thrives on the diversity of its collaborating disciplines and researchers. If we, as a society, really want to take on Big Questions, investments are needed across the board: in fundamental as well as applied sciences, in humanities and social sciences as much as in the fields of natural sciences, medicine, and engineering. Weaving research interests into collective challenges requires mutual curiosity and respect. Evidently, there are no guarantees for returns on investment, but science has always been more than a zero-sum game.

One of the most valuable outcomes of the NWA may pertain to the revitalizing of the art and science of asking questions. Keep in mind the words of James Stephens quoted above: 'We get wise by asking questions.' Asking questions, in other words, leads to wisdom, while finding answers leads to knowledge. This holds true for individuals as well as society at large.

Remember the secondary school girl and her question about the public debate on terrorism, cited in the beginning of this essay? My unwieldy proposal to cut her broad thesis into smaller ones did not resonate well. It took a few hours for the girl to reply to my email. 'Dear professor,' she wrote, 'thanks for your suggestion to come up with a smaller question. Unfortunately, I don't have time to do so, because my project needs to be finished the day after tomorrow. Could I please interview you so I can quote your answers to my questions about the power of mass media in my paper?'

Asking the right question in the right context is an important, yet increasingly neglected skill in the education of secondary school, undergraduate, and graduate students. Most schools' pedagogical plans are geared more towards conducting quizzes and finding answers – on the internet, most likely – than toward teaching a child how to articulate a sound question. Not surprisingly, it takes time to articulate a good question, a question that is both challenging and practicable, so the easy way out is to solicit answers from people who supposedly already 'have' that knowledge. Lack of time is also the reason why we increasingly cut parts of a graduate student's learning trajectory. In the Netherlands, unlike the US, the majority of graduate students are now recruited on the basis of project schemes prepared by their supervisor – obviously a successful fundraiser – so PhD students often do not get to articulate their own research questions. The quest for articulating a poignant yet 'researchable' thesis question is a fundamental skill that ought to be part of any student's education, most definitely a future professional. A system that does not allow for such an important part of a student's learning curve is in need of serious evaluation.

One of the interesting byproducts of the Dutch National Research Agenda released through its website is a lesson plan for secondary school students explaining the art and science of asking questions (Nationale Wetenschapsagenda voor scholieren, 2015). Indeed, asking researchable questions takes time and effort, and it is not an easy skill to teach, but it is an invaluable investment in the future wisdom of young people. If we want to stimulate a climate of infinite curiosity propelled by inquisitive minds, if we want our society to grow its potential for discovering the unpredictable, we need to invest in the art and science of asking questions. The Dutch National Research Agenda, if anything, turned out to be a national exercise in finding collaborative wisdom; this should be a tremendous gain for scientists, politicians, and citizens alike.

References

Government of the Netherlands, Ministry of Education (2014). The Dutch show the highest returns on EU research funding. Retrieved from www.government.nl/latest/news/2014/06/19/dutch-show-highest-returns-on-eu-research-funding

Hintum, Malou van, *Wat wil Nederland weten? De totstandkoming van de Nationale Wetenschapsagenda* (Masterdam: Nijgh & Van Ditmar, 2015)

Nationale Wetenschapsagenda (2015), https://vragen.wetenschapsagenda.nl/home

Nationale Wetenschapsagenda voor scholieren (2015), www.wetenschapsagenda.nl/scholieren/docenten/

Stephens, James, *Traditional Irish Fairy Tales* (New York: Dover publications, 1920)

About the author

José van Dijck is a professor of Media Studies at the University of Amsterdam. She has a PhD from the University of California, San Diego (USA). Her work covers a wide range of topics in media theory, media technologies, social media, television, and culture. She is currently President of the Royal Netherlands Academy of Arts and Sciences.

A Plurality of Voices

The Dutch National Research Agenda in Dispute

Henk Molenaar

In 2014, a new science policy framework was launched by the Dutch Ministry of Education, Culture and Science (*2025 Vision for Science*), heralding the development of a unifying agenda for research in the Netherlands. The agenda was to set out priorities and establish interrelationships between the research programmes of universities, research institutes, private sector companies, and other knowledge organisations. Ambitious guidelines and expectations were formulated. To mention only a few:

> The National Science Agenda is to be a 'co-creation' of researchers, scientists, the private sector, civil society, the government and other stakeholders. [...] The agenda will include a limited number of themes, selected on the basis of existing scientific strengths, societal challenges and economic opportunities. The research field as a whole will combine its strengths to achieve the greatest possible impact. [...] The National Science Agenda will appeal to the imagination; it will inspire and challenge both the research field and society itself to achieve momentous breakthroughs. It will create a better match between research on the one hand, and social and economic needs and opportunities on the other. It will clearly set out those areas in which the Netherlands is to stand out through truly excellent research. (ibid., p. 24)

In 2015, at the government's request, a coalition of umbrella organisations of the Dutch knowledge and innovation system (the so-called Knowledge Coalition) set out to develop and formulate the National Science Agenda. Amongst individual researchers these ideas and ambitions did not meet with universal enthusiasm. Quite a number of academics were sceptical and saw in the agenda the threat of a central top-down steering mechanism that would restrict their room for manoeuvre.

In the assignment letter to the Knowledge Coalition, the government added as further challenge the requirement to develop the agenda through an open process that would transcend existing institutional lines. The Knowledge Coalition met this particular challenge by organising a broad participatory bottom-up process. Anyone interested – whether universities, research institutes, civil society organisations, private companies, governmental organisations, or individual citizens – was given the opportunity to submit research questions. This approach met with a lot of enthusiasm and high expectations, but also with guarded reservations or scepticism. This time, some academics feared decentral bottom-up steering by the man in the street.

Almost 12,000 questions were submitted, far surpassing expectations. Juries composed of top researchers from all fields of the knowledge system grouped the questions into clusters and formulated overarching questions for each cluster. These were discussed in three conferences focusing on three different perspectives: Science for Science, Science for Society, and Science for Competitiveness. This was the basis for a further aggregation into a final number of 140 questions. In this way the questions submitted by the public were used as building blocks and sources of inspiration for the formulation of the 140 overarching questions which form the centrepiece of what has since been designated as the Dutch National Research Agenda.

This bottom-up process received much attention in the media and raised a lot of interest and enthusiasm. During the months in which the agenda was being developed, numerous bigger and smaller meetings, conferences and other forms of communication were organised, bringing science and the public into touch. This participatory approach contributed to enhanced public support for science and innovation. The involvement of the juries in constructing the agenda was another key success factor. Under the aegis of the Royal Netherlands Academy of Arts and Sciences and The Young Academy the juries were composed of eminent scientists. This manifestly resulted in growing support amongst the scientific community itself.

In the end, the agenda was well received and welcomed not only by the Dutch cabinet and the public at large, but also by the majority of the constituencies of the Knowledge Coalition. This was no mean accomplishment. At one stage or another, practically all parties in the knowledge and innovation landscape had expressed fears that the agenda would lead to a reallocation of research funding to their detriment. Diverging interests had to be aligned and expectations had to be managed.

Although the Dutch National Research Agenda does not aspire to be an all-inclusive agenda for science at large, its scope is nevertheless ambitious. The agenda focuses specifically on interdisciplinary and inter-sectoral challenges and as such stretches across the fields of science, technology, and innovation. It covers all sciences and academic fields of interest (natural sciences, life sciences, social sciences, humanities, and technological sciences) and embraces all types of research (basic research, strategic or policy-oriented research, applied and practice-oriented research). Consequently, it addresses and connects many different players within the Dutch knowledge and innovation system.[1]

In trying to cohere and integrate these various players and types of research under one single overarching agenda, it was necessary to take into account and come to terms with diverging perspectives, interests, ambitions and incentives. At times this met with resistance and led to fierce debates. Three disputes particularly emerged in this respect. The first was about the very notion of a research agenda and raised the question of whether a national research agenda could be to the detriment of curiosity-driven, unfettered research. The second dispute, closely related to the first one, was about the possible detrimental effects of an agenda focusing on a limited number of priority themes on the existing rich and multiform scholarly landscape. The third dispute started with questioning the wisdom of consulting the public at large in drawing up the national agenda and focused on issues of legitimacy and public support.

This essay reflects on these three issues by discussing the merits and shortcomings of some of the arguments raised. It also reflects on how choices made in developing the Dutch National Research Agenda relate to these disputes. In this way the essay situates the agenda in an analytical context that touches upon the nature of research and the sociology of science.

The research agenda in relation to unfettered research

A strong voice in the first dispute was that of academic proponents of curiosity-driven research. Amongst them the initiative to draw up a national research agenda met with scepticism or sometimes even outright hostility. They advocated unfettered research and experienced the national agenda as a threat, arguing that scientific advance cannot be steered, planned, or programmed. The free search of the human mind for new knowledge and insights, they argued, would only be hampered by an agenda. Indeed, the very concept of agenda-led research went against their grain.

---

[1] For that reason, in this essay the word 'science' is used in a broad sense covering all fields of systematic intellectual enquiry, including the social sciences and the humanities, and referring to research undertaken by all players within the national knowledge and innovation system, including private sector R&D.

Over time, many of those who were very outspoken and critical at first have become less fierce or even sympathetic to the Dutch National Research Agenda. But this does not hold for all of them and the agenda is bound to meet with this criticism time and again. No doubt, the persistence of this debate is related to the equally persistent drive of policymakers and funding agencies to earmark financial resources for specific social or economic purposes. In a context of fierce competition for scarce financial means, such choices unavoidably touch upon sensitive nerves and trigger emotional responses. There is good cause, therefore, to level-headedly scrutinize the reasoning behind these opposed lines of thinking. For is there truly a contradiction between agenda-led researches on the one hand and unfettered research on the other? It would seem that misconceptions are at play.

Free or unfettered research is sometimes referred to as 'blue skies research'. This metaphor indicates the mind transcending the limitations of earthly existence. On the ground, interests and agendas rule and mire the researcher in the mud of society. Such limitations hamper the curiosity-driven flight to great heights and new vistas. Unfortunately the metaphor confuses issues. It suggests that blue skies research is essentially basic research that needs to be distinguished from applied research driven by societal challenges and agendas. This confounds two different oppositions, one of basic versus applied research and another one of untied versus agenda-led research. Neither of these two distinctions is as pertinent as may seem.

Scientific progress shifts the frontiers of human knowledge. There is a widespread conviction that ground-breaking research is mostly basic in nature and that scientific breakthroughs only gradually find their way towards useful applications in society, sometimes after a delay of many decades. There are indeed many examples of such a course of affairs. But it is certainly not the only way in which new knowledge is created and utilized. Sometimes scientific breakthroughs – even fundamental paradigm shifts or the emergence of new disciplines – spring directly from social developments. One could think, for example, of the historical relation between bookkeeping and algebra (Crosby, 1997, pp. 204-220; Murray, 1978, p. 205; Soll, 2014, pp. 29-70) and between the insurance business and calculus (Tracy, 1985, pp. 212, 213).

The lineal knowledge chain that stretches from basic research via applied research to valorisation is only one of many patterns of knowledge creation and uptake. Sometimes basic breakthroughs find an immediate application in society. Applied or practice-oriented research, in turn, can give rise to new basic questions. Applied research is not to be confused with application. It is, indeed, research in which new knowledge is created, if only by merging or re-contextualizing existing knowledge. Basic and applied research can inform and enrich each other. The intellectual efforts involved are not at all different.

What about the other opposition, the one between untied and agenda-led research? This distinction also is not as clear-cut as it seems to be at first glance. How untied or free can scientific research really be? Within academia, research agendas abound. Every faculty, institute, or research group has an agenda. Such agendas focus research efforts, including curiosity-driven research that is not geared towards social challenges but towards questions that are relevant for the academic community itself. Such agendas do not come about haphazardly. They are based on foreseen and aspired scientific value.

Although disregarding economic value or social relevance, curiosity-driven research is not without direction or purpose. Next to economic values (profit, work, affluence), social values (well-being, social cohesion, peace and security) and ecological values (sustainability, biodiversity, conservation of nature), the creation of scientific values (insights and explanations, sense giving, knowledge as capability) should be recognized as a fully valid motive in itself. This motive informs agendas for curiosity-driven research.

How free can free science be in practice? Science is organized in disciplines. The designation 'discipline' is telling. Students are disciplined for years before being able and allowed to practise science. A researcher needs to learn and respect the – often tacit – codes of the discipline. Imparting knowledge to students is both a cognitive and a social initiation into the norms and customs of the field of study involved (Abma, 2011, p. 36; Kreber, 2009, pp. 19-31).

Freedom of research is relative in yet another sense. At the start of a scientific career many courses are open. However, a young academic is expected to abide by the agendas of supervisors, research schools, and the strategic plan of the institution he or she is affiliated with. At a later phase in life, a successful academic gains influence over such agendas through participation in committees and advisory boards. But by then his or her personal research efforts display path dependency based on the career path already travelled. A scientific career implies complying with many agendas, norms, and obligations. These limitations are accepted willingly and hence are not seen as constrictive. Freedom of research, therefore, is a matter of perception.

Of course, there is nothing against designating research geared towards the creation of scientific value as untied research, if this is meant to indicate that it is not motivated by social or economic goals. But the distinction with demand-led research should not be blown out of proportion. The distinction is not about research with and without a goal – the goals merely differ. Neither is it about research creating value versus research that is not creating value – it is only the types of value that differ.

Another argument sometimes raised against agenda-led research is the notion of serendipity. Scientific breakthroughs often arrive unexpectedly. They may come about as unintended side-effects of research into something else. This fact of life is brought forward as a plea for untied research. But such reasoning is misdirected. It implicitly assumes that untied research allows more space for serendipity than agenda-led research. However, serendipity needs no such podium and strikes at will. Serendipity occurs within every type of research, whether basic or innovation-oriented. Serendipity does not shy away from socially or economically motivated research agendas and is not lured by their absence.

Different types of research, therefore, are not all that dissimilar. The research agendas may differ, or the institutional settings or the conditions for research funding, but all types of research focus on creating new knowledge. A national research agenda should offer room for all these types of research. And this is precisely what the Dutch National Research Agenda does. It has been drawn up in such a way that it not only allows room for but also connects all types of research and research questions. The themes have been chosen and formulated to combine basic and applied research; connect curiosity-driven and innovation-oriented approaches, and bridge disciplines and sectors. The Dutch National Research Agenda transcends all such distinctions.

Disciplinary diversity and thematic focus

Another field of dispute encountered while developing the Dutch National Research Agenda was the fear that focusing on interdisciplinary themes would pose a threat to the wealth and diversity of the disciplinary landscape. Forcing research agendas into the mould of a limited number of thematic priorities, it was argued, would lead to a deterioration of the rich and multiform knowledge ecosystem. Small disciplines would run the risk of dwindling, facing the threat of extinction. As a result, the system at large could become less responsive to emerging possibilities and less resilient in dealing with external threats.

This dispute is even more intricate than the one about the threat of agendas to untied research, although there are certain similarities. It touches upon the very nature of the disciplinary organisation of academia and its state of tension with interdisciplinary research. In order to explain how the Dutch National Research Agenda dealt with this dilemma, it is necessary to delve into what disciplines are all about.

Disciplines do not reflect naturally given fields of reality. They are historical constructs created in the process of practising science (Abma, 2011, pp. 25-41; Rip, 2002, p. 125). They are created by people who identify objects of scientific inquiry, conceive of concepts and theories, conduct research, establish institutes for research and education, develop curricula, teach students, and create scientific journals. Disciplines are man-made ways of organising research and education. They grow into institutional frameworks that change over time and may differ from country to country. A discipline, therefore, is not a natural phenomenon but part of the sociology of science, useful for the organisation and reproduction of higher education and research.

Disciplines are both institutional and conceptual units (Becher, 1989, p. 20). The disciplinary framework cuts up reality into separate fields. The organisation of research and education in specialized disciplines allows for more in-depth knowledge creation. But this institutionalization of the process of knowledge creation also brings about constraints and lock-ins (Rip, 2002, p. 132). While reality is integrated and interdisciplinary in nature, disciplines compartmentalize research into silos and direct thinking into preset courses.

The social organisation of science is a play of inclusion and exclusion. Disciplinary knowledge is specialized, validated knowledge that is made available to some and not to others. Academics draw and sometimes dispute border lines with neighbouring disciplines. They demarcate their knowledge and insights from the ideas of others, especially from amateurs and lay practitioners. They claim exclusive authority in judging validity of knowledge in their field. And they constantly guard and strengthen the boundaries of their discipline (Abma, 2011, p. 31). This is particularly the case when career paths and other incentives are geared towards disciplinary excellence.

The partitioning of funding is another important underlying factor. A lot of public funding is earmarked for specific disciplines or groups of related disciplines. Moreover, financial resources are not equally divided over the various disciplines. Some disciplines have access to more funding windows than others, and the success rates in applying for research funding vary substantially from one field to another. This fuels competition between disciplines rather than cooperation.

As a result of all this, disciplines develop vested interests and mark identity by creating their own cultures with specific discourses, practices and rituals. This promotes their stability and reproduction, but may hamper effective interdisciplinary collaboration. Differences in tacit knowledge, norms and publication habits – not to mention diverging perspectives on ontology, epistemology, methodology, and pedagogy – are barriers to mutual understanding and adjustment (Donald, 2009, pp. 35-49). Such barriers may be even stronger when it comes to trans-discipline, cross-sector research involving non-scientific players (policymakers, enterprises, civil society, consumers, patients, and other end users) who stand to benefit or suffer from the outcomes of research. Divergent objectives and time frames can make such research collaboration quite a challenge (Molenaar, 2008, pp. 15-22).

Nevertheless, these forms of collaboration are urgently required. Humanity is beset by interrelated global challenges: wicked problems characterized by conflicting values, political pressure, moral confusion, and diverging economic interests. The complexity of these challenges can only be addressed through far-reaching systemic changes and transitions. Researching and meeting this complexity requires the involvement of many different parties and approaches, new connections and alliances. It calls for research integrating scientific and extra-scientific knowledge, experience and practice in problem-solving, taking the diversity of 'life-world' and scientific perceptions into account and linking abstract and case-specific knowledge (Edwards, 2011, pp. 7-16; Gibbons et al., 1994, pp. 1-17; Nowotny et al., 2001, pp. 48-55). Disciplinary silos do not easily allow for such partnerships (Kreber, 2009, pp. 19-31).

Still, the difference between disciplinary research on the one hand and interdisciplinary or trans-disciplinary research on the other should not be overstated. They have a common core situated in the very nature of knowledge creation. Knowledge creation is a social process, a collective endeavour. It requires the formation of an epistemic community, a community of peers understanding one another and collectively developing shared conceptual interests, lines of inquiry, and the practice of creating and validating new knowledge (Becher, 1989, p. 61).

Intriguingly, one can often observe interdisciplinary breakthroughs crystallizing into the birth of a new discipline (Rip, 2002, pp. 131-138). When effective, a newly formed epistemic community evolves most naturally into a discipline since this is the dominant way of organising knowledge production in modern societies. A successful epistemic community grows and diversifies. Different perspectives develop into specializations;

specializations grow into sub-disciplines. Informal lines of communication thicken into organised working arrangements. The new growing body of knowledge is introduced into higher education. A curriculum develops. A specialized journal is published. The institutionalization of the new discipline is underway.

In fact, most disciplines can trace their history back to an interdisciplinary or trans-disciplinary origin (Henry, 1997, p. 5). What presents an interdisciplinary theme today may grow into an academic discipline tomorrow. It is often argued that interdisciplinary research should be built on deeply rooted disciplinary work. That may be the case, but we must be cognizant that the deepest roots of disciplines are often interdisciplinary and cross-sector in nature.

Disciplinary and interdisciplinary research approaches, then, are not necessarily mutually exclusive or hostile to one another. Systemically, the one cannot exist without the other and for that reason academia needs to embrace both (Abma, 2011, p. 150; Rip, 2002, pp. 131-138). A too narrow focus on disciplinary excellence in splendid isolation may be fruitless in the long run. Disciplinary boundary work, therefore, should focus as much on building alliances and entering into partnerships as on guarding boundaries and defending territories. The sociology of science displays a geostrategic game played out in the landscape of academic 'tribes and territories' (Becher, 1989, p. 36).

A vibrant science, technology, and innovation system needs even more bridges and corridors, linking academia with extra-scientific players and sectors. It requires the possibility to reach out and connect beyond disciplinary and sector boundaries. The more diversified and multiform the knowledge landscape is, the greater the possibilities for teaming up and entering into new and unexpected alliances.

For this reason, a focus on a limited number of priority themes may not be in the interest of an effective national science and innovation system. Depending on the extent of the knowledge system, a too narrow specialization can indeed make the system vulnerable, less resilient, and less responsive to emerging threats and opportunities. In developing the Dutch National Research Agenda the Knowledge Coalition came to realise that in the comparatively highly developed and diversified Dutch knowledge and innovation landscape, building bridges across sectors and promoting inter- and trans-disciplinary research alliances is quite crucial and even a precondition for meaningful thematic prioritization.

The agenda has been drawn up in such a way that it can be used as an instrument for connecting different players in the knowledge and innovation system, for building new alliances, and for the joint programming of research. Practically all existing institutional research agendas to be encountered in the Dutch knowledge system have been identified and related to the questions of the Dutch National Research Agenda. Consequently, the agenda can be used as a map for exploring the science and innovation landscape to find potential partners. Furthermore, the instrument of 'routes' through the agenda has been developed. A route is a collection of agenda questions touching upon the various dimensions of a complex challenge, thus potentially linking a wide variety of parties who may be interested in teaming up to jointly meet this challenge.

The agenda has identified important issues and questions that call out for research and carry the potential for scientific breakthroughs in the years to come. These questions cover a wide range of topics and challenges. With the help of the route instrument, further prioritization of research themes to focus on can evolve through a bottom-up process addressing and tapping into the intellectual resources of the scientific community at large.

Legitimacy and public support

The third dispute started by questioning the approach followed in developing the Dutch National Research Agenda through consulting the public at large. As mentioned, this broad invitation to submit questions met with positive surprise and growing enthusiasm, but occasionally also with perplexity and confusion, both amongst researchers, practitioners, and the public at large. Although many acknowledged the importance of public support for research funding and recognized the bottom-up process of developing the national agenda as conducive in this respect, some wondered how an uninformed lay public could possibly set priorities for research.

This dispute both touches upon the self-image of academics (or researchers in general) and on the authority vested in science and the legitimacy thereof. It therefore merits a reflection on the nature of this authority and its role in society.

As mentioned, university education is a process of imparting guarded, specialized, validated knowledge. Not everyone is allowed to enter academia and benefit from this. There is selection at the gate. And not every student that enters succeeds in achieving the qualifications required to graduate. After successfully completing a university education, further steps await those who aspire to an academic career. In the process they may reach increasing levels of authority and legitimacy, each level with its own gatekeepers. Important steps on the ladder are marked with specific rituals. In this we can recognize the structure of the medieval guild system on which the earliest universities were modelled (Grant, 1996, pp. 34-39). It feeds a culture of exclusivity, which does not go unnoticed beyond the halls of academia.

Academia is one of three professional fields in which authority is marked by wearing a gown at official occasions, the others being the clergy and the judiciary. What these professional fields have in common is that they all set standards for and evaluate truth, validity, and righteousness and have the authority to disclose falsehood, errors, and transgressions. The gown symbolizes the legitimacy of the office bearer to wield this authority. But this legitimacy cannot be taken for granted. In recent decades the clergy has lost much of its former authority and the role of the judiciary is increasingly questioned and criticized nowadays.

So far academia has been spared such deterioration of authority (Hagendijk and Dijstelbloem, 2011, pp. 261-275). In fact, in the Netherlands there is substantial public support for and general interest in science. Mass media pay attention to scientific results and the public flocks to science festivals and science cafes. But continued trust in academia should not be counted on as a matter of course. Cases of plagiarism and falsified research badly hurt the reputation of science (Nelkin, 1996, pp. 114-122). Sometimes the outcomes of research are contested in society, because social impacts and ethical aspects have not adequately been taken into consideration (Kitcher, 2011, pp. 1-40). People increasingly use information on the internet to form their opinions. Medical practitioners are confronted with articulate and critical patients questioning their diagnosis. The gradual penetration of market principles in academia may further undermine the authority of science in the long run (Radder, 2009 and 2010).

This authority, therefore, needs to be carefully maintained and the legitimacy of scientists to wield this authority needs to be prudently justified and substantiated (Dijstelbloem et al., 2013, pp. 12-15). A most direct way to do so is to produce and widely communicate results. Scientific breakthroughs that help solve urgent problems and meet with priority needs are always welcomed. Many of the questions submitted for the Dutch National Research Agenda indeed relate to such problems and priorities. But this utilitarian function of science is certainly not the only important dimension. It is heartening to see that many individual citizens submitted questions that have no direct social relevance or economic value but touch upon the fundamentals of life and the universe.

Providing meaningful answers to such basic questions must be recognized as meeting an important social need. There is a lot of interest in the origins of the universe, the earth, life, and mankind. But also in a wider sense many look to science for sense giving and for interpreting man's position in the world. The social sciences and humanities play an important role in providing conceptual and normative frames for debating complex challenges (Radder, 2014, p. 5). This dimension is often overlooked by politicians and science policymakers, but it is crucial for public support and the authority of science. There is wide support for these roles of science, although the public may not at all be aware through which mechanisms the related research is funded.

In this context pleas by academics for additional funding for free and untied research can be inexpedient and counterproductive. This is the case when such advocacy takes on a denunciative form, denying policymakers, funding agencies, or the public at large the right or competence to earmark funding. Enhanced sensitivity for how such advocacy may be perceived would be wise. As it happens, the plea for free and untied research perfectly matches the self-interest of researchers since it enlarges the scope for the research they choose to conduct. When the argument is added that researchers themselves are best qualified to assess which research should be funded (the Haldane principle), the perception of the pursuit of self-interest grows stronger. This unavoidably touches a nerve with funders and policymakers. Advocacy that does not take their role, responsibility and mindset into account may do more harm than good.

Individual academics would do best to emanate the importance of all types of research. And the best way to do this is to actively reach out to the general public and to specific audiences. In fact, consciously building new audiences could be considered a task of growing importance for academia and non-academic research institutes (Dijstelbloem, 2014, p. 49). To the extent possible, such audiences can be involved in drawing up research agendas and in the research process itself. This will enhance trust, enlarge public support, and legitimize the authority of the sciences.

Concluding remarks

Looking back upon these three disputes one can observe that the process of creating the Dutch National Research Agenda has resulted in a meeting of minds, both within the scientific community itself and beyond. New partnerships already emerged during the process of harvesting questions and formulating the agenda. Many more partnerships are likely to follow in the years to come. The disputes have raised the level of mutual understanding across the partitioning of sectors and disciplines. But these disputes are far from over, as the various contributions to this volume attest, and this is as should be. Reasoning will continue to be deepened and new insights will emerge. The Dutch National Research Agenda will continue to spark off debates on science policy and on the sociology and philosophy of science.

The broad public consultation that was the start of formulating the Dutch National Research Agenda and, subsequently, the many meetings that were organised during the process to bring the public into contact with scientists have contributed to enhanced interest in and public support for science. In this, the process of developing the agenda met with some of the principles of responsible research and innovation as advocated by the European Commission (2013).

Equally important was the involvement of the scientific community in processing the many questions submitted by the public and in formulating the overarching questions, thus embracing and making the most of this hugely varied input and taking seriously the bottom-up process of formulating the agenda. In doing this they discarded or weakened many of the misgivings and doubts they experienced at the start, such as discussed above, and developed a measure of ownership over the Dutch National Research Agenda.

In the long run, the success and impact of the Dutch National Research Agenda require not only sustained interest and involvement of the public, but also engagement with and commitment to the agenda by the scientific community itself. This calls for continued dialogue and communication. The present volume aspires to this effect.

References

Abma, Ruud, *Over de grenzen van disciplines. Plaatsbepaling van de sociale wetenschappen* (Nijmegen: Van Tilt, 2011)

Becher, Tony, *Academic tribes and territories. Intellectual enquiry and the culture of disciplines* (Buckingham: Open University Press, 1989)

Crosby, Alfred W., *The Measure of Reality. Quantification and Western Society, 1250-1600* (Cambridge University Press, 1977)

Dijstelbloem, Huub, 'Science in a not so well-ordered society: A pragmatic critique of procedural political theories of science and democracy', *Krisis. Journal for contemporary philosophy*, 1, 2014

Dijstelbloem, Huub, Frank Huisman, Frank Miedema, and Wijnand Mijnhardt, *Waarom de wetenschap niet werkt zoals het moet, en wat daar aan te doen is* (Science in transition. Position paper: 2013, www.scienceintransition.nl/over-science-in-transition/position-paper)

Donald, Janet Gail, 'The Commons. Disciplinary and Interdisciplinary Encounters', in *The University and its Disciplines. Teaching and Learning Within and Beyond Disciplinary Boundaries*, edited by Carolin Kreber (New York: Routledge, 2009), pp. 35-49

Edwards, M., *Thick problems and thin solutions: How NGOs can bridge the gap* (The Hague: Hivos, 2011)

European Commission, *Options for Strengthening Responsible Research and Innovation*, Directorate-General for Research and Innovation (Brussels: European Commission, 2013)

Gibbons, Michael, Limoges, H., Nowotny, S., Schwartzman, S., Scott, P., and Trow, M., *The New Production of Knowledge: The Dynamics of Science and Research in Contemporary Societies* (London: Sage, 1994)

Grant, Edward, *The Foundations of Modern Science in the Middle Ages. Their Religious, Institutional, and Intellectual Contexts* (Cambridge: Cambridge University Press, 1996)

Hagendijk, Rob and Huub Dijstelbloem, 'Omgaan met onzekerheid in wetenschap, politiek en samenleving', in *Onzekerheid Troef. Het betwiste gezag van de wetenschap*, edited by Huub Dijstelbloem and Rob Hagendijk (Amsterdam: Van Gennep, 2011), pp. 262-275

Henry, John, *The Scientific Revolution and the Origins of Modern Science* (Basingstoke: Palgrave Macmillan, 1997)

Kitcher, Philip, *Science in a Democratic Society* (New York: Prometheus Books, 2011)

Kreber, Carolin, 'The Modern Research University and its Disciplines. The Interplay between Contextual and Context-transcendent Influences on Teaching, in *The University and its Disciplines. Teaching and Learning Within and Beyond Disciplinary Boundaries*, edited by Carolin Kreber (New York: Routledge, 2009), pp. 19-31

Ministry of Education, Culture and Science of the Government of the Netherlands, *2025 Vision for Science: choices for the future* (The Hague, 2014, www.government.nl/documents/reports/2014/12/08/2025-vision-for-science-choices-for-the-future)

Molenaar, Henk, 'Introduction and synthesis: Towards a new understanding of research for development', in *Knowledge on the Move. Emerging Agendas for Development-oriented Research*, ed. by Henk Molenaar, Louk Box, and Rutger Engelhard (Leiden: International Development Publishers, 2008), pp. 1-29

Murray, Alexander, *Reason and Society in the Middle Ages* (Oxford: Clarendon Press, 1978)

Nelkin, Dorothy, 'The Science Wars: Responses to a Marriage Failed', in *Science Wars*, edited by Andrew Ross (Durham, North Carolina: Duke University Press, 1996)

Nowotny, Helga, Scott, P., and Gibbons, M., *Rethinking Science: Knowledge in an Age of Uncertainty*, (Cambridge: Polity Press, 2001)

Radder, Hans, 'Hoe herwin je de ziel van de wetenschap? Academisch onderzoek en universitaire kenniseconomie', *De Academische Boekengids*, 75, 2009, pp. 8-12

Radder, Hans, 'The commodification of academic research', In *The commodification of academic research: science and the modern university*, edited by H. Radder (Pittsburgh: University of Pittsburgh Press, 2010), pp. 1-23

Radder, Hans, *Waartoe Wetenschap? Over haar filosofische rechtvaardiging en maatschappelijke legitimering*, Rede uitgesproken bij het afscheid als bijzonder hoogleraar in de filosofie van wetenschap en technologie aan de Vrije Universiteit Amsterdam (Amsterdam: Hans Radder, 2014)

Rip, Arie, 'Science for the 21st Century', in *The Future of the Sciences and Humanities*, edited by Peter Tindemans, Alexander Verrijn-Stuart, and Rob Visser (Amsterdam: Amsterdam University Press, 2002)

Soll, Jacob, *The Reckoning: Financial Accountability and the Rise and Fall of Nations* (New York, Basic Books, 2014)

Tracy, James D., *A Financial Revolution in the Habsburg Netherlands. Renten and Renteniers in the County of Holland 1515-1565* (Berkeley: University of California Press, 1985)

National Research Agendas

An International Comparison

Wim de Haas

Introduction

In November 2015 the Dutch National Research Agenda was published. This agenda describes in 140 overarching questions the major scientific challenges for the future. The agenda was written at the request of the Dutch Ministry of Education. The idea for such an agenda followed from the National Science Vision 2025. According to this Science Vision, the Dutch National Research Agenda should play a steering role in Dutch science policy.

In general, public investment in research is justified from the perspective of the contribution of scientific research to social, cultural, and scientific development, as well as to innovation. In Dutch science policy, as in many other countries, the latter 'innovation argument' has become increasingly important, and the economic crisis of recent years has put even more emphasis on it. Research agendas play a particular role within the scope of science policy instruments. They tend towards thematic prioritization of investments and of other policy instruments.

In this context, the aim of this essay is to explore and compare some aspects of national research agendas in order to examine the position of the first Dutch National Research Agenda. First, the essay considers the policy context of national research agendas. Second, fifteen countries are examined to determine whether or not a country has a national research agenda. Third, looking at countries with national research agendas, these agendas are compared and the character of the agendas and the themes that are prioritized discussed. In addition, the essay describes the process of development of the agendas and some aspects of implementation. Finally, the Dutch National Research Agenda is compared to the other agendas.

Science and innovation policy as context for national research agendas

In many countries science policies show three consistent transitions. First, a transition from direct funding by the government to funding through a system of calls and tenders, performed by national research councils or comparable institutions. In line with this, a second transition occurred: a turn from a supply- to demand-driven knowledge 'production'. Thirdly, theme-oriented science policies evolved in addition to general science policies. The emergence of the policy instrument of a national research agenda is consistent with this third transition. In a comparative study of six European countries, Lepori et al. (2007) show three comparable developments: an increase in project funding, a differentiation of instruments and an increase in thematic prioritization.

This general shift towards thematic prioritization started in the 1970s as a result of social motives, especially the need to control technological developments. This was motivated by negatively perceived effects of technology and science on social well-being and on the environment. From the 1990s, the motives for thematic prioritization shifted towards the need to innovate, which became stronger in the economic crisis at the beginning of the 21st century.

In many countries not only science policy but also technology and innovation policies are important for research funding. In these policies innovation is generally considered as technological innovation, but social innovation receives greater attention. The general trend can be characterized as a transition from industrial support to innovation policy. In the nineties, many countries supported increases in funding, emphasizing that innovation policy should focus more on key industrial sectors than on lagging or newly developing sectors. This was inspired by the ideas of the economist Michael Porter (1990) and is sometimes characterized as: 'backing winners', as opposed to 'backing challengers', i.e. targeting promising new sectors, or 'backing losers', i.e. supporting companies in trouble. In the Netherlands, for instance, innovation policies are now partly aimed at nine key industrial sectors: knowledge-intensive sectors with a substantial contribution to export.

Against the background of this general development of thematic prioritization, there are some interesting differences between countries. Especially small countries seem to specialize in specific research areas. Soete et al. (2012, p 16) provide an overview of differences in innovation policy (Table 1). They show which countries focus more on proven strengths, such as the Netherlands, and which countries invest more in new dynamics. Israel and the United States are examples of the latter. In addition, supporting 'specific targets' can be distinguished from investing in 'broad absorption'. Broad absorption is the ability to incorporate information and knowledge and to transform it into insights or judgements that enable new innovations (WRR, 2008). The absorption capacity of a national economic system can be enhanced, for instance, by investments in education. Some countries, such as Finland, combine the latter with a backing winners approach, while China and Germany combine it with a focus on new dynamics.

The table below from Soete et al. (2012, p 16) is just a rough characterization. For example, the Netherlands is characterized in the table as aiming at specific targets, but also has a general tax reduction policy for R&D investments by private companies, in which more public money is invested than in the 'specific targets' of the 'backing winners' policy (Jacobs and Velzing, 2013).

Table 1 Characterisation of innovation policy in several countries

Innovation policy aim	Specific targets	Broad absorption
Proven strength (*Backing winners*)	The Netherlands: top sectors Switzerland	Sweden(?), Finland Denmark, Japan (?)
New dynamics (*Backing challengers*)	Israel USA	China Germany (?)

Source: Soete et al. (2012)

Three discourses as a context for research agendas

Research agendas emerge as an important policy tool in the briefly described developments in science, innovation, and technology policies. The specific role of the agenda depends on the dominant concepts and theories about the mechanisms that connect research and innovation. Herein three discourses are manifested (De Haas et al. 2014).

The first is a discourse on stimulating *general conditions* for innovation, such as tax reduction for R&D, and enhancing the absorption capacity, for instance by investment in education. In this discourse, research and innovation are characterized as evolutionary processes that can only be stimulated by general measures supporting the conditions for innovation. The role of thematic research agendas is a general exploration of new topics rather than a steering instrument. A national thematic agenda is mainly an analytical and explorative instrument.

The second discourse is focused on the idea that explicit *thematic choices* must precede a successful relationship between research and innovation, implying that thematic innovation policy works. The concept of 'Backing winners', focusing attention and resources on existing and proven strengths, is part of this discourse. Thematic research agendas play an important role in this discourse as an instrument of prioritization.

The third discourse is based on the assumption that *networks* between companies, researchers, and governments are essential for a fruitful relationship between research and innovation. In this discourse, agendas are considered less important than the exchange of ideas and knowledge. This networking mechanism is in essence related to specific problem areas or sectors. This discourse, in the Netherlands known as the 'golden triangle', manifested itself successfully in the Dutch agricultural sector (OECD, 2015, p 136). In this discourse a national research agenda represents agreements made by network partners.

Additional analyses

The cooperation between companies, universities and research institutes, and governments is sometimes also described as 'Triple Helix' (Etzkowitz, 1998; Etzkowitz and Leydesdorff, 2000). This is an analytical model that combines two points of view. The first is an institutional viewpoint that focuses on actors and their cooperation. The second is a social-evolutionary point of view which distinguishes between the production of prosperity, production of innovations, and normative control. The Triple Helix model is then more than a practical policy choice for better cooperation between government, industry, and knowledge institutions. This model is extended by others by inclusion of NGOs. Carayannis and Campbell (2009) further extend the Triple Helix model with a cultural dimension. This refers to the mix of actors who operate in the media, in creative industries, the arts, the culture sector, etc., also called the 'creative class' (Florida, 2004).

With reference to the second discourse, research practice does not always react as intended, according to research by Van den Besselaar en Horlings (2011). They showed that the concentration of research resources to key thematic areas (*'sleutelgebieden'*) in former Dutch Science Policy had a limited effect on the number of publications in these areas. This was possibly caused by the absorption capacity of the Dutch research system. Researchers are effective in articulating the big goals of the government in concrete terms, as indicated in a recent study by Bos (2016).

In Dutch innovation policy ('top sector policy') all three discourses are apparent (De Haas et al. 2014). In short, this policy combines general instruments from the first discourse with the choice for top sectors from the second discourse. Both are held together by the rhetorical use of the 'golden triangle' metaphor from the third discourse. The Dutch top sector policy is therefore an example of what Hajer (1993) calls discourse coalition, in which even opposing discourses have found a way to cooperate.

National research agendas

This section provides a brief analysis of national research agendas or other kinds of national thematic research prioritization in fifteen countries with well-developed science and innovation policies (Table 2).

Table 2   National research prioritization: characterization for fifteen countries

Country	National thematic research prioritization	Characterization	Cycle (years)
European countries			
Denmark	Yes	Thematic research agenda: *The Research2015 Catalogue*. Priorities: 21 themes in six fields.	4
France	Yes	*Strategic Agenda for Research, Technology Transfer and Innovation*. Aimed at improving the research system. Prioritization around nine major social challenges.	'Will be regularly revised'
Germany	Yes	*High-Tech Strategy*. Broad agenda for mid-term innovation policy. Technological and social innovation. Aimed at system improvement and strategic prioritization. Six thematic priorities.	*
Ireland	Yes	*National Research prioritization*. 14 priority areas around which future investment in publicly-performed research should be based. Aimed at commercial outcomes and sustainable businesses and jobs.	5
Italy	Yes	*National Research Plan*. Main target-setting instrument for research investments in Italy. One of the main targets will be reinforcing the strategy of international research. For basic, applied, and industrial-related research. Seven scientific macro-areas.	3
The Netherlands	Yes	*Dutch National Research Agenda*. 140 questions divided into 16 'routes'.	7
Poland	Yes	*National Research Programme*. Strategic Research directions for the long-term directions. Seven priorities.	10-15
Sweden	No	No explicit national research policy or agenda. No overall vision for the whole system.	n/a
Switzerland	Yes	Periodically renewed set of *National Research Programmes*. Chosen by the national government; substantial bottom-up influence.	2-3
United Kingdom	No	No national strategic prioritization. Seven Research Councils have own strategies and research prioritization.	n/a

Country	National thematic research prioritization	Characterization	Cycle (years)
Non-European countries			
Australia	Yes	*National Innovation and Science Agenda* aimed at improving the research system in general. Followed by the *Science and Research Priorities Australia* with nine priorities.	2
Japan	Yes	*Comprehensive strategy on science, technology, and innovation* as a long-term vision for 2030 to achieve an ideal economic society. Includes the whole picture of science, technology, and innovation policies and action programme. Five priorities, each worked out in 2-5 challenges.	*
Korea	Yes	*Vision 2030*. Five-year Basic Plan for Science and Technology. Regularly updated. A comprehensive long-term strategy to transform Korea into a fully advanced country. Selection of 30 priorities in four fields and 120 strategic technologies.	5
Singapore	Yes	*Research, Innovation, Enterprise 2020 Plan*. Integrated technology and science prioritization to improve health care, boost the economy, and create jobs. Major shifts to capture more value from research. Four strategic technology domains.	5
USA	No	No national thematic research agenda. Large national research initiatives on certain topics, in some cases on specific laws.	n/a

* not indicated

While it is difficult to take all the specific circumstances in each country into account, a number of interesting points can be noted. Most of the fifteen surveyed countries do have some kind of national thematic research prioritization. In most cases, this prioritization is meant to be renewed every three to five years.

Particularly in Asian countries, the national research agenda is closely linked to the overall economic and innovation policy. In Korea and Singapore, this mid-term innovation policy is regularly updated. Japan has a regularly updated mid-term agenda, but also formulated a long-term strategy. These countries show the relevance of a thematic research agenda towards a leap forward in innovation (OECD, 2009).

Furthermore, it appears from this overview that especially smaller European countries such as Denmark, the Netherlands, and Switzerland have chosen national thematic prioritization of research. This may indicate that smaller countries feel the need to make specific choices or find specific niches to compete. Nonetheless, even countries without a national agenda, such as the United States or Sweden, do have an extensive and proven system of prioritization at the level of sectors, disciplinary science foundations, or otherwise.

Priorities

This section discusses the content of the national research agendas; which themes are prioritized in the agendas? Table 3 shows an overview of the thematic prioritization.

Most national research agendas have a rather broad scope, which means that they do not focus only on technology and innovation, but on the entire range of social issues. A specific feature of the Dutch Research Agenda is that it is made up of questions and 'routes' connecting these questions. Asian countries show a strong focus on technology and innovation, embedded in a strategy for general economic and social development.

A solid comparison is difficult because the agendas' priorities are formulated at different levels of aggregation. Nevertheless, the priorities show a large overlap. Many topics appear on several agendas, for instance energy, sustainability, food, and various health-related topics.

Table 3 Prioritized themes in national research agendas (in italics: themes mentioned five times or more; bold: some notable research themes, for various reasons)

Country	Prioritized research themes
Denmark	Fields: *Energy*, climate and environment; Production and technology; *Health* and prevention; Innovation and competitiveness; Knowledge and education; **People and social design**.
France	Resource management and adaptation to climate change; Clean, secure, and efficient *energy*; Stimulating industrial renewal; *Health* and well-being; *Food* safety and the demographic challenge; *Sustainable* mobility and *urban* systems; Information and communication society; Innovative, integrating, and adaptive societies; A spatial aspiration for Europe.
Germany	Digital economy and society; *Sustainable* economy and *energy;* Innovative workplace; *Healthy* living; Intelligent mobility; Civil security.

Country	Prioritized research themes
Ireland	Future Networks & Communications, Data Analytics, Management, Security & Privacy, Digital Platforms, Content & Applications, Connected *Health* and Independent Living, Medical Devices, Diagnostics, Therapeutics: Synthesis, Formulation, Processing and Drug Delivery, *Food* for Health, *Sustainable Food* Production and Processing, Marine Renewable *Energy*, Smart Grids & Smart *Cities*, Manufacturing Competitiveness, Processing Technologies and Novel Materials, Innovation in Services and Business Processes.
Italy	Scientific macro-areas: *Food, Energy*, Society, Nanotechnology, Mobility, *Health*, Safety.
The Netherlands	Sixteen 'routes' through 140 questions: Personalised *medicine*; Regenerative *medicine*; *Health* care research; The origin of life; Building blocks of matter and fundaments of space and time; Resilient and meaningful societies; Between conflict and cooperation; **Brain, cognition, and behaviour**; Using big data responsibly; Smart industry; Smart, liveable *cities*; Circular economy and resource efficiency; *Sustainable* production of safe and healthy *food*; **Arts**; Quality of the environment; Logistics and transport. The agenda is open to other routes.
Poland	New energy-related technologies; Diseases, new *medicine* and regenerative *medicine*; Advanced information, telecommunication and megatronics technologies; New Materials; Natural environment, agriculture, and forestry; Poland's social and economic development; State security.
Switzerland	Big data, Smarter *Health* Care, Antimicrobial Resistance, Managing *Energy* Consumption, *Energy* Turnaround, Healthy Nutrition and *Sustainable* Food Production, *Sustainable* Use of Soil, **End of Life**, Resource Wood, New *Urban* Quality, Nanomaterials, Regenerative *Medicine*, Smart Materials, **Gender Equality**.
Australia	*Food*, Soil, and Water, Transport, Cybersecurity, *Energy*, Resources, Advanced Manufacturing, Environmental Change, *Health*.
Japan	Clean and economic *energy* system; *Healthy* and active ageing society; Next generation infrastructure; Regional revitalization; **Recovery and revitalization from the great East Japan earthquake**.
Korea	Traditional priorities: several industries. New priorities: the green economy, the creative economy.
Singapore	Strategic Technology Domains: Advanced Manufacturing and Engineering; *Health* and Biomedical Sciences; Services and Digital Economy; *Urban* Solutions and *Sustainability*

Process and implementation

In developing a research agenda, three different methods are recognized (Table 4, second column).
1   The first addresses a large number of parties including citizens. The Dutch Research Agenda is a good example of this. It started with an invitation to citizens and organisations to submit questions to science.

2 The second method consults various parties outside the government, but is restricted to parties from science and industry. The Irish Research Prioritization is a good example of this.
3 The third method incorporates the agenda as part of a regular policy process. Asian countries often follow this procedure to develop their research agendas.

For the implementation of national research agendas, two models are distinguished (Table 4, third column).

A In one model, the agenda is included in a regular update of the research priorities. Next, these priorities are worked into programmes by research councils.
B In the other model, the calendar plays a role in the renewal of research and innovation policy: in some cases as the first time for a new regular prioritization process, in other cases as part of an overall renewal of the research or innovation system.

Table 4 Process of development and implementation of national research agendas

Country	Process (methods 1, 2, 3)	Implementation (models A, B)
Denmark	Mapping of research needs by a literature scan, a broad public internet hearing, input from the ministries. Expert panels delivered themes. The selection of final priorities was discussed with organisations, ministries, and research councils. (1)	Implementation by the national research council. Inspiration for universities. (A)
France	Close consultation with the scientific community, social and economic partners, the relevant ministries, and local authorities. (2)	Will be implemented through multi-year contracts concluded with research institutions, higher education institutions, the National Research Agency's (ANR) planning department, and other public research funding agencies. (B)
Germany	The High-Tech Strategy has been developed by the government in close cooperation with representatives from industry and science. (2)	Federal projects; coordination (departments, *Länder*); impact analysis. Public involvement in the innovation process; social innovation. (B)
Ireland	Initial deliberations with science organisations. Six expert groups. Steering group made final proposal for the government. (2)	Implementation is the responsibility of the government departments and agencies. (B)

Country	Process (methods 1, 2, 3)	Implementation (models A, B)
Italy	Normal ministerial policy process. (3)	Distribution of resources among the funds of science foundation. (A)
Netherlands	Broad bottom-up process, selection and combination by expert panels (juries); final proposal by a steering committee representing all Dutch science organisations. (1)	Government intends to use it for prioritization in research policy and agreements with universities. (B)
Poland	Draft prepared by the Scientific Policy Committee and discussed with ministries, councils, and agencies. The choice of strategic research priorities was made with the participation of 'distinguished representatives of various communities', especially researchers. (1 / 2)	Worked out by the National Centre for Research and Development into strategic programmes. (A)
Switzerland	Interested parties (federal offices, research institutes and groups, and individual persons) can submit topics and priorities for National Research Programmes. The Federal Council judges these and makes the final selection. (1)	The Federal Council defines the budgets and commissions the Swiss National Science Foundation SNSF to implement the NRPs. (A)
Australia	Chief Scientist in consultation with researchers, industry leaders, and government representatives. (2)	Over time, the priorities will result in an increased proportion of public investment in science and research going to areas of critical need and national importance. (B)
Japan	Priorities are determined along institutional lines. (3)	Reallocation of resources for research by the government from a long-term agenda. (B)
Korea	Regular updates by taskforce of representatives of technology and engineering organisations, research institutes, and universities. (3)	Large role for the government to adapt the science and technology system and allocate resources to priorities. (A)
Singapore	Developed by the National Research Foundation: a department under the Prime Minister's Office. Advised by a committee with representatives from industries and universities. (3)	Worked out in programmes by the National Research Foundation. Emphasis on public-private partnerships. (A)

Dutch National Research Agenda compared to other agendas

Most national research agendas are part of an existing research or innovation policy cycle: the agendas represent choices and are meant to allocate research funds. Two aspects distinguish the Dutch National Research

Agenda from most other agendas. The first is the open call to anyone to submit questions. The second is the choice to describe a number of routes through the landscape of submitted questions instead of a prioritization of some themes. The reasons behind both aspects are possibly found in the traditional autonomy of universities in the Netherlands and the preference for extensive consultation and consensus in Dutch administrative culture. Moreover, the agenda is the result of cooperation between science organisations with, at some points, different interests. These aspects encourage an agenda that transcends interests rather than an agenda based on strong choices.

How does the meaning of the Dutch National Research Agenda compare to the agendas in other countries? In this respect, three roles of a research agenda can be distinguished, in three keywords: lobby, policy preparation, and science communication. In the short term, the Dutch agenda functions as a kind of lobby instrument; the agenda plays a role in the debate on the amount of research funding for the next years, using the bottom-up character of the agenda and the consensus between knowledge organisations as arguments. This role is not found in the agendas of other countries. The policy preparation role is relevant for the medium term. According to the '2025 Vision for Science' of the Dutch government, the agenda will play a role as a seven-year prioritization instrument in the regular update of science policy. This role of the Dutch Agenda corresponds fully with that of the other agendas. The science communication role is relevant for the long term, allowing the agenda to play a role as a continuous articulation of public questions to science. This role is also found in some other agendas; in Switzerland and Denmark, the public has a role in bringing up new ideas and topics. Perhaps this last role is the most challenging, as it can be of great significance for the public commitment to science in the long term.

Conclusions

In this essay, some aspects (context, character, themes, process, implementation) of national research agendas in fifteen countries were compared in order to examine the position of the Dutch National Research Agenda.

Thematic prioritization of research, by means of an agenda, is a general trend that can be observed in most countries. This fits in with a discourse on science policy that emphasizes applying focus. Thematic prioritization is also related to the increased importance of innovation as grounds for science policy. In some of the fifteen countries, research agendas are part of a regularly adjusted national innovation policy, while in other countries the agenda has a broader scope than just innovation. The Dutch National Research agenda belongs to the latter group.

The themes mentioned in the examined research agendas are largely comparable. Many countries prioritize themes such as energy, sustainability, and health issues. With regard to the preparation of the agendas, two approaches are observed; some countries prepare the agenda as a process between governments, companies, and researchers, while other countries have tried to incorporate citizens in the preparation process. In this respect, the Dutch agenda is unique. It started with a broad invitation to citizens and organisations to submit their questions to science. This approach has the potential to be used for a continuous articulation of research questions from the public, which could be of great importance for the public support of science.

References

Bos, C., *Articulation. How societal goals matter in nanotechnology.* PhD thesis (Utrecht: University Utrecht, 2016)

Carayannis, Elias G. and David F.J. Campbell, '"Mode 3" and "Quadruple Helix": Toward a 21st Century Fractal Innovation Ecosystem', *International Journal of Technology Management* 46 (3),, 2009, pp. 201-234

De Haas, Wim, Kristof van Assche, Marcel Pleijte, and Trond Selnes, *Gouden Driehoek? Discoursanalyse van het topsectorenbeleid* (Wageningen: Alterra Wageningen UR, 2014)

Florida, R.L., *Cities and the Creative Class* (New York: Routledge, 2005)

Hajer, Maarten, 'Discourse Coalitions and the Institutionalization of Practice: the Case of Acid Rain in Britain', in *The Argumentative Turn in Policy Analysis and Planning,* edited by Frank Fisher and John Forrester (Durham and London, Duke University Press, 1993)

Jacobs, Danny and Evert-Jan Velzing, *Innovatie- en industriebeleid in het begin van de 21ste eeuw – topsectoren, fiscale regelingen en een techniekpact (Innovation and industry policy at the beginning of the 21st century – Top sectors, tax regulation, and a national technology agreement)* (Hoofddorp: Stichting Industriebeleid en Communicatie, 2013)

Etzkowitz, H., and L. Leydesdorff, 'The Triple Helix as a Model for Innovation Studies', *Science & Public Policy* 25, 3, 1998, pp. 195-203

Etzkowitz, H., and L. Leydesdorff, 'The dynamics of innovation: from National Systems and "Mode2" to a Triple Helix of university–industry–government relations', *Research Policy* 29, 2000, pp. 109-123

Lepori, Benedetto, Peter van den Besselaar, Michael Dinges, Bianca Potì, Emanuela Reale, Stig Slipersæter, Jean Thèves, and Barend van der Meulen, 'Comparing the evolution of national research policies: what patterns of change?', *Science and Public Policy*, 34, 6, 2007, pp. 372-388

OECD, *Reviews of Innovation Policy: Korea* (Paris: OECD, 2009)

OECD, *Innovation, Agricultural Productivity and Sustainability in the Netherlands* (Paris: OECD Food and Agricultural Reviews, OECD Publishing, 2015)

Porter, M.E., *The Competitive Advantage of Nations* (New York: Free Press, 1990)

Soete, Luc, et al. *De wedloop om kennis. De kennissamenleving in internationaal perspectief* (The Hague: Adviesraad voor Wetenschaps- en Technologiebeleid, 2012)

Van den Besselaar, P.A.A., and E. Horlings, *Focus en massa in het wetenschappelijk onderzoek: de Nederlandse onderzoeksportfolio in internationaal perspectief* (The Hague, Rathenau Instituut, 2015)

Wetenschappelijke Raad voor het Regeringsbeleid (Scientific Council for Government Policy, WRR), *Innovatie vernieuwd: opening in viervoud* (Amsterdam: Amsterdam University Press, 2008)

The Role of Universities of Applied Sciences in Implementing the Dutch National Research Agenda

Daan Andriessen and Marieke Schuurmans

Introduction

The academic landscape of the Netherlands is divided into two types of universities: research universities and universities of applied sciences (UAS). The universities of applied sciences outnumber the research universities by 37 to 14. They host almost twice as many students as research universities, 446,500 versus 250,000. Both universities offer bachelor and master programmes. Universities of applied sciences provide higher professional education, preparing students for specific professions. The programmes offered tend to be more practice-oriented than programmes offered by research universities. Since 1986, research has been a designated task of universities of applied sciences (Knoers, 1995), but it has only grown into a serious activity since 2001, when the first professors were officially installed. Research can be conducted in collaboration with research universities, but this is not compulsory. If a research results in a PhD thesis, however, collaboration with a research university professor is obligatory. Professors at universities of applied sciences are not assigned with the *ius promovendi*, the legal position to award the degree of PhD.

In this chapter we discuss the possible contribution of UAS to the implementation of the National Research Agenda (*Nationale Wetenschapsagenda*, or NWA). The Netherlands Association of Universities of Applied Sciences has been a member of the knowledge coalition and the steering committee of the Dutch National Research Agenda (Hintum, 2015). Member universities organised ten sessions on various topics resulting in a total of 150 questions submitted to the NWA. In our opinion the universities of applied sciences can also play an important role in the implementation of the NWA. In this essay we shall explore this role. We'll start with providing an overview of the development of the research role of the universities of applied sciences. Then we will reflect on three key issues that touch upon the implementation of the NWA:

1   research programming versus the need for free research;

2   the legitimacy of research in politics and society;
3   the need for focus and clustering.

We will discuss each of these three issues from the perspective of universities of applied sciences whose core strength lies in doing research in close collaboration with professional practitioners. Finally, we will describe three prerequisites for maximizing the contribution of UAS to the implementation of the Dutch National Research Agenda.

Practice-oriented research at universities of applied sciences

Table 1 shows some key figures on research in universities of applied sciences. In this section we describe the nature of research at universities of applied sciences. Since 2001, the nature of this research and the differences with research undertaken in research universities have been strongly debated. The Advisory Board on Science and Technology (in Dutch *Adviesraad voor wetenschap, technologie en innovatie:* AWTI), an influential advisory council of the Dutch government, argued that research in universities of applied sciences should be referred to as 'design and development' (Adviesraad voor Wetenschaps- en technologiebeleid, 2001). According to the advisory board, the task of contributing to science is the exclusive right of research universities and therefore the term 'research' should be reserved for them. However, in 2010, in a new law governing the higher education sector, Dutch Parliament decided to use both the terms 'research' and 'development' for universities of applied sciences, thereby indicating that their role is both to develop new knowledge and solve practical problems.

Table 1   Key figures for Dutch universities of applied science (2014)

Number of universities	37
Number of students	446,500
Core tasks	Education, research, and development
Type of research	Practice-oriented research
Number of professors	592 (65% male, 35% female)
Fte of professors	361 FTEs
Number of researchers	3,548
Fte of researchers	1,037 FTEs
Researchers in a PhD trajectory	865

Traditionally, Dutch universities of applied sciences have strong relationships with practice. Most of them have evolved from educational programmes initiated by trade organisations and similar interest groups (Van Bemmel, 2006). Educational programmes are developed in cooperation with practice and students often do internships at a company or institution. Research conducted at universities of applied sciences has a similar orientation towards practical work and innovation. In 2007, the Association of Universities of Applied Sciences described research at universities of applied sciences as having roots in professional practice and generating knowledge for direct use in professional practice. The research is often multidisciplinary in nature and is based on co-creation with professional practitioners. It is scientifically robust and has strong connections with both education and professional practice.

Some still feel an urge to differentiate research conducted at universities of applied sciences from research at research universities. At one point the term 'applied research' was chosen to make that distinction (HBO-raad, 2000). The downside of this particular term is that it directly refers to the distinction between basic research and applied research first used by Vannevar Bush (1945, p.18): 'Basic research is performed without thought of practical ends. It results in general knowledge and an understanding of nature and its laws. This general knowledge provides the means of answering a large number of important problems, though it may not give a complete specific answer to any one of them. The function of applied research is to provide such complete answers'. This distinction is based on a linear model of innovation in which new knowledge is exclusively generated by basic research undertaken by (natural) scientists that then gets applied to practice through applied research. In applied research no new knowledge is created. Seventy years after Bush this linear view of innovation is outdated (Vasbinder & Groen, 2002). The application of basic research outcomes is not the only source from which innovations spring, nor is the development of new knowledge the exclusive domain of basic research. For that reason, we oppose the use of the term 'applied research' as a label for the research conducted at our universities. The Association of Universities of Applied Sciences agrees and has decided to use the term 'practice-oriented research'. Unfortunately the legacy of Bush has such a strong foothold in the Anglo-Saxon world that our universities are known in English as universities of applied science.

The work of Gibbons et al. (1994) can help to further clarify practice-based research at UAS. They make a distinction between mode 1 and mode 2 knowledge production, where mode 1 is traditional ivory tower research and mode 2 multidisciplinary research conducted in close cooperation with practitioners. Gibbons et al. claim mode 2 to be a new mode of knowledge production, emerging in the middle of the 20th century and displaying five characteristics: context application, transdisciplinarity, heterogeneous practices, reflexivity, and novel forms of quality control. Research at Dutch universities of applied sciences shows many mode 2 research characteristics, although not all five are equally applicable in all cases.

Practice-oriented research is not the exclusive prerogative of universities of applied sciences. In our view it is not very fruitful to make a strong distinction between the types of research conducted by the two types of universities. In both universities one can come across research that has mode 2 characteristics. In contrast, at Dutch universities of applied sciences, one will not encounter pure basic research. All research is based on questions derived from practice and produces new knowledge that is applicable in practice. In Dutch universities of applied sciences, there is no room for questions that solely spring from the personal curiosity of the researcher or from the blanks in scientific theory.

The core strength of Dutch universities of applied sciences lies in the close relationships with professional practice. All research is based on problems or opportunities that arise in the society, in the daily practice of companies, hospitals, schools, welfare institutions and the like. The research questions are often explicitly articulated together with those working in the field. Examples include research into ways that small and medium-sized companies can benefit from biopolymers and smart materials (Saxion); research into ways that journalists can make use of infographics (University of Applied Sciences Utrecht); research on how to introduce student teachers in conducting and using research (Fontys) and research guiding optimal use of instruments by healthcare professionals (Hogeschool Zuyd).

In fields like social work it is common to involve practitioners in the design and execution of the research. Sometimes a research project is not merely used to generate knowledge but also to implement change within an organisation. Approaches such as action research (Kemmis & McTaggart, 2000) or design-based research are common (Van Aken, 2011). In many cases the result of practice-oriented research is knowledge that can be used directly in local situations, designated by Argyris (1996) as 'actionable knowledge'. This is in contrast with explanatory sciences whose mission is primarily to describe, explain, or predict (Van Aken, 2005). However, proper practice-oriented research aims not only at local problem-solving but also at generating knowledge that has wider implications than a single context. This occasionally remains a challenge.

Research results are disseminated through various means. Peer-reviewed journals are not the primary focus of the Dutch universities of applied sciences. Nevertheless, publishing in such journals is encouraged since peer reviews increase the quality of the research and help to strengthen the relationship with research universities. Research is disseminated through professional journals, reports, books, websites, and by creating products for practice. An important instrument for dissemination is the research process itself. By conducting the research in close cooperation with practitioners, knowledge is disseminated both explicitly and implicitly. Training or empowering the professional in the field may be an explicit goal of the research. Last but not least, the collaboration with students and their teachers within the research projects provides a strong vehicle for early dissemination of research results.

Another core strength of research at universities of applied sciences is that science is not the only source of knowledge in research projects. Because of the close relationships with professionals in the field, knowledge of professionals and clients or patients can be included. This knowledge is made explicit, evaluated, and tested.

The research effort by Dutch universities of applied sciences has grown considerably since the start. In 2001, the first professor was appointed and in 2014, there were 592 professors (361 FTEs) (Vereniging Hogescholen, 2016) of which 35% female. For most of them, the professorship at the university of applied sciences is a part-time job. Many combine it with a position in a company, research university, or other institution. Most professors have their own research group consisting of teachers in the role of researcher. On average a research group consists of 6 researchers, each having 0.3 FTEs to do research, leading to a sum total of 3,548 researchers and 1,037 FTEs, of which 17% have a PhD ibid.).

The Dutch universities of applied sciences have the ambition that 10% of their lecturers will be trained at doctorate level. The majority of the growth comes from teachers following a PhD trajectory at a research university; 865 in total in 2014. The Dutch Ministry of Education, Culture and Science strongly advocates the value of practice-oriented research at universities of applied sciences. It has set the ambition to increase the volume of professors to 580 FTEs by 2024. With the current part-time factor this means an increase to 950 professors (Ministerie van Onderwijs, 2015).

The current €171 million of research funding derives from three sources. 63% is so-called first-stream funding by the Ministry. The remaining 37% is second- and third-stream funding, including funding by a dedicated fund for practice-oriented research at universities of applied sciences (€18 million) and the European Union (€5 million).

Research planning versus the need for free research

The first of the three core issues that are central in this chapter is the role of research planning. What is the origin of research, where do questions stem from and how do researchers assess the importance of these questions in UAS? The Dutch universities of applied sciences feel that free research is the task of research universities. Their own strength lies in the close connection with professional practice. Their research programmes are built on the explicit needs and wishes of professional partners (and on their educational programmes) on the one hand, and the expertise of the professors they attract on the other.

All individual research projects start with a problem from professional practice. All grant funding parties of practice-oriented research judge the relevance of research and the explicit articulation of the research question from a practice perspective. For professors coming from research universities it is sometimes challenging to develop research questions on the basis of professional practitioners' problems. For some professors, however, the practice perspective is the very reason they switched position from a research university to a university of applied science. They feel that the focus in research universities on publishing in high-ranked scientific journals hampers doing useful and relevant research.

The close collaboration between researchers, teachers, and practitioners in practice-oriented research, sometimes even in the form of co-creation, stimulates adoption of findings and shortens the time lag between knowledge creation and knowledge use. Research and dissemination often go hand in hand. Involving practitioners in choosing research subjects, formulating research questions, conducting research, and disseminating results can be a huge learning experience for them. At the same time this collaboration makes it possible for research to gather professional knowledge, smart solutions, tips and tricks that have been developed in practice, and to research the effectiveness of this type of knowledge and make it available for other practitioners to use. In this way, practice is not only a source of data but a source of valuable knowledge as well.

To conclude, within universities of applied sciences no tension is felt between research planning and free research. Therefore, the NWA is seen by many as an opportunity and not as a threat. Many questions in the NWA have a practice focus. Questions like No. 15: 'How can we create more sustainable food-producing systems?', or No. 10: 'How can we make buildings and infrastructure safer, more sustainable and less costly using new materials, technologies and processes?' address societal problems that practitioners struggle with. Therefore, contributing to research based on the NWA will not be too difficult for UAS researchers. They are used to planning their research from a user perspective. Some NWA questions or 'routes' fit very well with the profile and research portfolio of various UAS. For example, one of the routes through the 140 questions of the Dutch National Research Agenda is about smart, liveable cities. The research programme of the University of Applied sciences Utrecht focuses on improving the quality of living in urban environments.

The legitimacy of research in politics and society

The second core issue in this chapter is about the legitimacy of the research. What is the legitimacy of the research conducted at universities of applied sciences? In as little as fifteen years, universities of applied sciences have developed a research function that has gained trust among politicians and is valued by society. An important factor is that the research questions are close to daily life and are understandable for all. In addition, the practical relevance of the research becomes increasingly evident and parties start to appreciate the work done. For example, in 2014 over 4,600 SMEs were involved in projects funded by the NRPO-SIA, a dedicated fund for practice-oriented research at universities of applied sciences.

However, the legitimacy of research conducted by universities of applied sciences is still fragile in the eyes of research universities and the scientific community. Research universities have been sceptical from the beginning. Questions were raised regarding the critical mass, the academic climate, the rigour of the methodology and the expected quality of results. One reason is that the growing role of research at universities of applied sciences is seen as a threat to the ambitions of research universities. Research funding in the Netherlands does not grow proportionally with the number of parties doing research.

Another reason is that the quality of research within the universities of applied sciences is far less transparent compared to research universities. Research universities have stronger mechanisms in place to ensure quality and to calibrate quality standards within specific areas of research. They have, for example, procedures for consultation of sister faculties when appointing professorships. There are strong research communities in which professors know each other as a result of peer reviews of PhD theses, papers, and grant proposals. In contrast, the appointments of professors within universities of applied sciences are local procedures that vary between individual universities and in which peers within the field do not play a specific role. The research communities are less strong and professors in the same field sometimes do not know each other personally. Professors at universities of applied sciences do not have *ius promovendi* and are in many cases less involved in the international research community. Their research programmes are not subject to regular calibration with standards in the field. Many professors at research universities are not aware of the work of their colleagues at universities of applied sciences and vice versa.

Research at universities of applied sciences is much less frequently subjected to peer review. International scientific publications are not the key output. Publications are aimed at dissemination to the field of professional practitioners. Furthermore, the organisation and governance structures within universities of applied sciences are not yet fully adapted to the research responsibilities. Research experience is frequently lacking in boards of directors or amongst directors of institutes and other leadership positions. This sometimes results in policies that hamper the work of researchers or lack a focus on research quality. To strengthen this focus, the Association of Universities of Applied Sciences has recently developed a policy demanding the use of explicit quality criteria to review and improve research (Vereniging Hogescholen, 2015). This is a first step; however the effect largely depends on the extent to which the criteria will be applied. A non-binding policy will not enhance the general quality of research from universities of applied sciences.

To conclude, the political and societal legitimacy of research at UAS is growing but the scientific legitimacy needs further improvement. For universities of applied sciences to play an effective role in implementing the Dutch National Research Agenda, it is necessary to improve the visibility of the professors and their work. Moreover, to sustain political and societal legitimacy and at the same time gain the respect of research universities, quality of research is crucial and transparency of practice-oriented methodologies is required. For this a more obligatory quality policy is required. In December 2014, the Association of Professors at universities of applied sciences was formed.[1] The purpose of the association is to promote the quality and visibility of practice-oriented research.

---

[1] www.lectoren.nl

The need for focus and clustering

The third core issue is the need for focus and clustering. How do universities of applied sciences deal with this? In the first decade of research at the universities of applied sciences, research programming was done by individual professors. There was not much cooperation between professors within the universities, let alone between universities. However, in the last five years much progress has been made. A big step was the creation of Centres of Expertise in which universities of applied sciences collaborate with practitioners to close the gap between research, education, and practice.

After ten years of experimentation, most universities of applied sciences have now decided to cluster their professors in knowledge centres that focus on particular subjects. The purpose of clustering research is to increase focus and combine research capacity in order to improve research quality and impact. The positioning of these centres within the university varies. Some are tied to educational faculties and led by the faculty dean, others are positioned close to the board of directors of the university.

Many universities of applied sciences are in the process of developing research programmes based on societal themes. For example, University of Applied Sciences Utrecht focuses on improving the quality of living in urban environments, and Saxion focuses on Living Technology. However, the way in which these programmes actually steer research is not yet fully crystallized. Several models coexist but we will mention only three. First, in some cases research programming is merely a language game in which prioritizing is nothing more than semantics. Second, sometimes research programming takes the form of identifying focal points for which additional resources beyond base-funding are available. And third, and this is the most extreme form of steering, a centralized body within the university decides on research projects to be undertaken. To conclude, at many universities of applied sciences research programming is still very much a paper exercise. Individual professors find it hard to give up their autonomy in deciding what research to undertake. A certain level of autonomy is important, but some coordination of research efforts is needed to improve excellence and impact, and financial incentives can help. The NWA can be a useful tool to stimulate the debate, to develop connections between research programmes, and to strengthen ties with research universities. Moreover, working within collaborative programmes between different universities provides a strong mechanism not only to improve quality but also to reduce research waste. There are many causes of research waste, ranging from poor research programming to the choice of methodology or a lack of consistency between research phases. The NWA can help to create unifying pathways from basic to applied science and vice versa, thereby reducing research waste.

Contributing to the Dutch National Research Agenda

In our opinion, the Dutch universities of applied sciences are very well-positioned to contribute to the implementation of the Dutch National Research Agenda. The focus of UAS on practice-oriented research and their strong network in professional practice will ensure that the Dutch National Research Agenda truly contributes to society. Many questions posed within the NWA have a practice-oriented dimension and demand clear-cut answers that can change the way we build our cities, organise our healthcare system, and deal with migration.

Implementation of the NWA requires strong collaboration between all parties. In our view, three prerequisites are essential to optimize this collaboration, each involving a changing view on research and innovation:
1 transition from a linear to a cyclical and network view;
2 transition from a monodisciplinary to a transdisciplinary view;
3 transition from a hierarchical to a non-hierarchical view.

These three transitions are briefly expanded on below.

From a linear to a cyclical and network view

As described earlier, innovation is not a linear process from basic research through applied research to new products and services. It is an iterative process in which many parties are involved, each bringing their particular strengths to the table (Vasbinder & Groen, 2002). In cyclical innovation, basic research is very much needed. However, this basic research can be supplemented with more practice-oriented research that studies practical problems and can inform basic research about instruments, applications, important factors that have been overlooked, implementation issues and the like. It can also be complemented with entrepreneurial activities that involve experimentation and risk-taking. Crucial to success is the creation of networks that can facilitate this collaboration. Early crossovers between basic and practice-oriented research can catalyse and speed up findings in both. In a network view on innovation it is not useful to create a strict division of labour between research universities and universities of applied science but to profit from the strengths of both.

The Dutch National Research Agenda can be a strong catalyst for the creation of these networks. Many questions in the Dutch National Research Agenda have both a basic and practice-oriented component. Many include both descriptive and explanatory questions as well as design questions. For example, question No. 5: 'What is the role of micro-organisms in eco systems and how can these be used to improve health and the environment?' includes both an explanatory question that requires basic research and a design question that requires practice-oriented research. In order to fulfil this catalyst role, much more effort must be put into identifying parties involved in each of the 140 questions and in validating the information entered in the database.

From a monodisciplinary to transdisciplinary view

Solving the complex problems of today's society requires knowledge from various disciplines. Not only by looking at these problems from different perspectives (multidisciplinary research), but also by creating new knowledge through combining various disciplines (interdisciplinary research) and by thinking from each other's perspectives and disciplines (transdisciplinary research) (Rosenfield, 1992). One of the challenges for universities of applied sciences is to incorporate more of the tools, methods, and theories of basic research into their work. The scientific merit of practice-based research can be improved. At the same time the challenge for many research universities is to incorporate a practice-oriented perspective into their work and make more use of research methodologies that have been developed with this in mind.

From a hierarchical to a non-hierarchical view

Transdisciplinary research requires close collaboration between disciplines and between research universities and universities of applied sciences. For this to happen, we need to leave behind the tendency to think in terms of a hierarchy of forms of knowledge or research. The Netherlands is praised for its non-hierarchical culture and some ascribe the success of Dutch science to the fact that in Dutch culture researchers dare to oppose their professors and debate among equals is common. Yet, in our experience, thinking in terms of a hierarchy is still very much present when it comes to the relative positions of research universities and universities of applied sciences.

The idea that research at research universities is of higher quality or more profound hampers a closer collaboration between all universities. The fact that universities of applied sciences don't have *ius promovendi* creates a hierarchy and dependency between the two types of universities that impedes integration of knowledge, ideas, and methods. At the same time it hampers the calibration of quality standards across the knowledge system and the full recognition of each other's work. Competition for research funding hinders the close collaboration that is needed to implement the Dutch National Research Agenda. To realise the ambition of answering all questions incentives for a change of attitude and behaviour and for collaboration across the entire university landscape are recommended.

References

Adviesraad voor Wetenschaps- en technologiebeleid, *Hogeschool van Kennis* (The Hague: Adviesraad voor Wetenschaps- en Technologiebeleid, 2001)

Argyris, C., 'Actionable Knowledge: Design Causality in the Service of Consequential Theory', *Journal of Applied Behavioral Science*, 32 (4), 1996, pp. 390-406.

Bush, V., *Science, The Endless Frontier* (Washington, DC: National Science Foundation, 1945)

Gibbons, M., C. Limoges, H. Nowotny, P.S.S. Schwartzman, P. Scott, and M. Trow, *The New Production of Knowledge – The Dynamics of Science and Research in Contemporary Societies* (London: Sage Publications, 1994)

HBO-raad, 'Hogescholen benoemen lectoren' (Position paper), *Hogeschoolberichten 241*, 2000. Retrieved from www.vereniginghogescholen.nl

Hintum, M. van, *Wat wil Nederland weten?* (Amsterdam: Nijgh & Van Ditmar, 2015)

Kemmis, S., and R. McTaggart, 'Participatory action research', in *Handbook of qualitative research,* edited by N.K. Denzin and Y. S. Lincoln, 2nd Vol. (London: Sage Publications, 2000), pp. 567-605

Knoers, A.M.P., *Onderwijs in de Europese Unie – Het onderwijs in Nederland* (Heerlen: Open Universiteit, 1995)

Ministerie van Onderwijs, C. en W., *Investeringsagenda van de Strategische Agenda Hoger Onderwijs en Onderzoek 2015-2015* (The Hague: Ministerie van Onderwijs, Cultuur en Wetenschappen, 2015)

Rosenfield, P.L., 'The potential of transdisciplinary research for sustaining and extending linkages between the health and social sciences', *Social Science and Medicine*, 35(11), 1992, pp. 1343-1357.

Van Aken, J.E., 'Management research as a design science: Articulating the research products of Mode 2 knowledge production in management', *British Journal of Management*, 16(1), 2005, pp. 19-36.

Van Aken, J.E., 'Ontwerpgericht wetenschappelijk onderzoek', in *Handboek Ontwerpgericht Wetenschappelijk Onderzoekonderzoek,* edited by J.E. Van Aken and D. Andriessen (The Hague: Boom Lemma Uitgevers, 2011), pp. 25-39

Van Bemmel, A., *Hogescholen en hbo in historisch perspectief* (The Hague: HBO-Raad, 2006)

Vasbinder, J.W., and T. Groen, *Tussen Kennis en profijt; Hoe onze samenleving veel meer kan halen uit kennis* (Warnsveld: Prisma & Partners, 2002)

Vereniging Hogescholen, *Brancheprotocol Kwaliteitszorg Onderzoek* (The Hague, 2015)

Vereniging Hogescholen, *Praktijkgericht onderzoek; Factsheet 2012-2014* (The Hague, 2016)

Steering Scientific Research and Reaping its Benefits

Reflections on Dutch Science Policy

Coenraad Krijger[1] and Maarten Prak[2]

Introduction

Much is expected from scientific research. Governments hope to see solutions for complex policy challenges. Industries and businesses hope for innovations supporting their competitive edge. Society as a whole hopes for a deeper understanding of a complex world and safeguards for welfare and well-being. Last but not least, the community of scholars hopes to be able to understand, reflect, and explore. In the past, society was patiently waiting for results to emerge, trusting that clever scientists would be making new discoveries. Nowadays research is considered too important to be left to its own dynamics and, for that matter, to scientists themselves. The potential of science is to be reinforced, its impact and benefits channelled and increased. As a result, science policies have emerged that also raised an interest, first among politicians and policymakers, and more recently also among the general public.

Science policy is, however, confronted with a fundamental problem. By definition, the results of research projects cannot be predicted. If they could be, the research would be futile. Because of the huge expenses involved – in the Netherlands currently an estimated 4.5-5 billion euros annually[3] – the government and other policy institutions nonetheless hope to steer these investments towards useful, efficient, and targeted outcomes. In other words, science policy hopes to increase the predictability of results. With this in mind, policies are formulated with objectives ranging from nurturing, facilitating, and supporting scientific research to steering, streamlining, and orchestrating its topics and processes. Typically, national science policies

---

[1] *Coenraad Krijger is director of IUCN NL, a non-profit organisation for nature conservation. At the time of writing he was director of policy development at the Netherlands Organisation for Scientific Research (NWO). This essay reflects his personal views.*
[2] *Maarten Prak is Professor of Social and Economic History at the Department of History and Art History at Utrecht University and Chair of the Humanities Board of the Netherlands Organisation for Scientific Research.*
[3] Totale Investeringen in Wetenschap en Innovatie 2014-2020. Rathenau Instituut 2016.

demonstrate a policy mix of measures that can best be explained as a sum of decisions accumulating over time and taken more or less independently at different levels, by successive governments and boards.

In this essay we want to discuss several assumptions underlying science policies and reflect on how, given what is known about systems for research funding and about the effects of policy interventions, the Dutch National Research Agenda (*Nationale Wetenschapsagenda*, or NWA) is embedded in the Dutch academic system. In the course of this paper we will argue that it is not so much the policies themselves that guide or determine the success of science, but the way they add up and shape the environment in which scientists and scholars do the actual work. In order to steer them and reap their benefits, policies should fit the inherent dynamics of the scientific system.

Types of research and funding systems

Is there something like an ideal science policy? And if so, what would it look like? As a start, it should deal, in an effective way, with the diversity in research approaches. A first observation must be that science policy has a tendency to ignore this fundamental point. A lot of the debate on science policy, and also much of the academic literature that contributes to this debate, demonstrates a remarkably narrow view of research. It holds up the natural sciences model as the prototype, or tends to simply reduce scientific research to the natural and life sciences, without further ado. Scientific (including technical) research is usually theory-driven, produces its own data in secluded environments (the 'laboratory'), and results in statements that are thought to be universally applicable. However, this model applies only in certain parts of the research world, and not even in all of the sciences. Mathematics, for example, is not a laboratory science; much technical and engineering research is trial and error rather than theoretically framed. The model is to a large extent not applicable outside the natural and life sciences. Anthropologists, historians, law scholars, or philosophers produce statements and explanations that are context-specific, based on unique observations collected 'in the wild', and they are therefore sceptical of the broad generalisations that we call 'theory'. In disciplines like economics, linguistics, psychology, and sociology we find both types of research. These disciplines are therefore characterized by fierce struggles over methodology, with the 'laboratory' type of scholars berating the poor methodology of their colleagues, who in turn point out that the laboratory produces results that could be irrelevant to the real world. A successful research policy should take these variations into account, or must result in a withering of significant branches of research. This would be a problem not only for the world of research, but for society itself. The success of innovations depends, by and large, as much on its social context as on technology as such (Volberda and Van den Bosch, 2013, p. 44).

As there are different types of research, so we find different types of research funding. Three broad categories have been distinguished, each with its own characteristics (Lepori, 2011, pp. 362-4). In the United States, around eighty percent of public research funding is provided through project subsidies. Ministries, research councils, technological agencies, and other public bodies hand out money to research groups and individuals who have to submit funding applications. YThis is essentially a market model for stimulating research, which has the effect that relatively large amounts of funding go to a limited number of stakeholders ('Matthew effect'). This model still requires a larger pool of competitors, to avoid oligopolistic effects when a small number of research groups stifle the positive effects of competition. It is therefore less suited to smaller systems, where the number of competitors almost by definition will also be smaller. No surprise then that European countries use a different system that channels much of the research funding through universities. In these countries, project funding covers typically between twenty and forty percent of public research. In the Netherlands, the national research council NWO is responsible for about 20 percent of publicly funded research (Koier et al., 2016, p. 38).[4] This structure of funding seems to be especially well-suited for higher education systems with even distributions of research facilities; Switzerland, Norway, Finland, and indeed the Netherlands are usually cited as examples. A third model is the vertically integrated system that was popular in Central and Eastern Europe during the communist era, but also in post-war Southern Europe, including France. In this system, a single, large research facility or institution, usually the Academy of Sciences or national research institution such as the CNRS in France or the CSIC in Spain, is charged with research, while the universities are primarily educational institutions.

In past decades, mixes of these three have evolved, at least in the larger countries in Europe, towards a mix of all three systems, and other European countries are also evolving towards such a mix. In the Western European model, the objective of project funding is to dynamise the research system with the help of strategic incentives. It would therefore be short-sighted to create a funding structure that serves one single purpose.

---

4   In terms of absolute volume; the indirect effect of NWO funding is considered to be bigger when the total costs associated with the funded research are taken into account.

Impact of academic research

In recent years, the 'impact' of science, in addition to its 'excellence', has become an important issue for policymakers. Politicians argue that excellent science cannot be a goal for its own sake. They want to know how the society they represent can benefit from scientific discoveries and, where possible, increase these benefits. Here, too, policies tend to start from unduly simplistic assumptions. The 'impact' debate has had a tendency so far to focus on economic benefits, and there again on the direct profits that might be reaped from research. However understandable against the backdrop of the recent economic crises, this is an unnecessarily narrow, and in our view ultimately counterproductive, interpretation of the 'impact' of scientific research. Astronomy is not focused on immediate economic benefits but helps to locate us, as humans, in the wider universe. And yet, it gives rise to high-tech innovations taken up by industries. Understanding the marvels and complexity of the living world and its diversity brings us much more than ideas on how to explore or preserve it. The popularity of nature documentaries and their development to include the latest scientific insights can illustrate this. Likewise, understanding how Rembrandt produced his masterpieces does not lead to immediate economic benefits, but enriches our understanding of a unique artistic achievement and can serve as an inspiration for future generations. To be sure, a Rembrandt exhibition, or the Rijksmuseum's presence in Amsterdam, has major economic benefits, and the research underpinning the museum's presentations contributes to those benefits. It is, however, difficult to calculate how cost-effective art history is as a discipline. Likewise, legal scholarship underpins the justice system, sociological research addresses issues with the integration of migrants and refugees, while pedagogy helps to improve our school system. It would be difficult to deny their importance, even if it remains impossible, and is perhaps even morally wrong, to ascribe a precise monetary value to their contribution.

Having said this, scientific research has proven to be fundamental for our economic prosperity, and increasingly so. Even if we stick to economic benefits for the sake of the argument, the literature distinguishes six different dimensions of research impact (Salter and Martin, 2001, pp. 518, 520-26).

a   Increasing the stock of useful knowledge: especially publications create opportunities to access new knowledge that firms and organisations can apply in their work processes.
b   Training skilled graduates: possibly the most important effect of research is the production of a skilled workforce, to be employed by

firms and organisations. Advances in human capital formation are generally seen as the single most important growth factor in advanced economies.

c Creating new instrumentation and methodologies: this is an important result of government-funded (rather than industry-funded) research, but its impact is difficult to measure.
d Forming networks and stimulating knowledge interactions: firms located close to major centres of academic research benefit more from that research than those located at a distance, as a result of social and professional interactions between their employees and academics.
e Increasing the capacity for problem-solving. A Yale survey of 650 R&D directors showed that fundamental scientific knowledge impacts 'through influencing the general understandings and techniques that industrial scientists and engineers, particularly those whose training is recent, bring to their jobs' (Klevorick et al., 1995, quoted in Salter and Martin, 2001, p. 525). This was confirmed by a similar European survey.
f Creating new firms: the most famous example is, of course, the impact of Stanford University on the creation of Silicon Valley in California, but on the East Coast, MIT has had a similar though perhaps less dramatic effect, and in the Netherlands we can see the same effect around Eindhoven University, for example.

Impact, too, is therefore poorly served by a one-size-fits-all approach. Furthermore, it should be noted that these effects are the result of general research, not driven by specific training, and one might even argue that their effect could be jeopardized by too much specialization in a handful of areas. We should also keep in mind that the economic, let alone social, political, or cultural, impact of research will always be difficult to forecast due to the inherently unpredictable character of research (Dasgupta and David, 1994, p. 490).

Innovative and routine research

Research is enamoured with innovation. The 'first' discovery of a particle, effect, or other breakthrough is rewarded with Nobel Prizes, Field Medals, and similar distinctions. Much research policy is likewise obsessed with the prizes that seem to signal success as a precursor of future breakthrough discoveries. One often-heard criticism of such policies is that 'it would not have made a difference for Einstein'. There are two reasons why this is not a very strong argument. The first is that it seems really difficult to predict the emergence of Einsteins. So far, a strong correlation has been established between excellence and funding levels. The lavishly endowed Oxbridge colleges in the UK, and Ivy League universities in the US, seem to do really well in terms of Nobel Prizes, because they have the facilities to attract the very best scholars. This simultaneously improves those top-level institutions and impoverishes the rest. It is, in other words, a typical example of 'beggar thy neighbour' and not a policy that could be repeated by many other countries. This implies that countries where such world-class research facilities are not widely available (basically everybody else) have to design different research policies, adequate to their own specific research environments.

The truth of the matter is – the second problem with the Einstein argument – that most researchers are not Einsteins, and that most research is not ground-breaking. And this is just as well, because next to novel ideas we also need more precise knowledge about how things really work. In fact, we need solid testing and replication of research findings, most certainly if the research is to have 'impact'. Routine research, or basic research, in other words, is as necessary as ground-breaking research, especially if it is taking local circumstances into account. University rankings identify world-class institutions, but not necessarily world-class systems. Currently (2015), the Times Higher Education world university ranking classifies 17 US universities in the top 25. The United States, in other words, completely dominates. Classifications that rank countries according to the number of universities in the world's top 200, and take the size of their economies (GDP) into account, give us a very different picture. In 2012, when these figures were compiled, among the ten highest ranked countries, six were located in Europe (Times Higher Education World University Rankings, 2011-12, p. 17). The Netherlands was classified second, after Hong Kong, while the UK, Switzerland, Sweden, Ireland, and Denmark also made the grade, in that order. This ranking seems to suggest that the current policy mix in the Netherlands works rather well, and that perhaps it would be easier to spoil the 'magic potion' than to improve it.

The literature about the organisational and institutional preconditions for creativity suggests factors that stimulate creative research. These include opportunities for multiple interactions with colleagues, staff mobility, communication across disciplinary boundaries, and leadership by scholars who are themselves still active in research. As far as organisational aspects are concerned, a survey among 185 American and European experts in human genetics and nanoscience and nanotechnology also underlines several points that are important for research policy (Heinze et al., 2009, pp. 616-19).

a   Agenda-setting: broadly defined problems and long-term targets allow focus and freedom at the same time.
b   'Complementary variety': research units will become more creative when they are regularly exposed to ideas from groups working in adjacent fields, with different skills and methodologies.
c   Flexible funds: creative groups indicate that they benefit from funding that allows them to follow their ideas, instead of being constrained by very specific budgets. Block grants, rather than project grants, help to create this sort of flexibility.
d   Job mobility: grant schemes could encourage this by creating the right conditions.
e   Funding supportive of risks: moving from one field to another takes a lot of time. The best part of five years may be required to build up a credible publication portfolio in that new field to help attract new funding. Funding agencies do not usually support this type of move. Moreover, they require well-defined targets, at the expense of exploratory, open-ended projects.

These observations suggest that a mixture of funding tools will best support a healthy research biotope (Laudel, 2006, p. 384). That mixture should include targeted and open-ended funding, allow spending flexibility, and stimulate cross-disciplinary interactions and researcher mobility.

'Blue skies' and 'brown earth' research

What drives scientific and scholarly research? Or rather, what inspires scientists and scholars to employ their talents and creativity to advance our understanding and search for solutions? In policy circles a distinction is often made between the various purposes of scientific research in an attempt to influence and steer the course of scientific progress. Various classifications and concepts circulate, such as 'fundamental', or 'blue skies', research that is supposed to be driven by scientific excellence, contrasted with 'use-inspired', 'applied' or what we might call 'brown-earth' research, which is supposed to be more focused on 'useful' results for the real world. Often these types of scientific research are depicted in contrast to one another, competing for funds and attention.

The current European science policy employs a threefold classification: Science for Science, Science for Society, and Science for Competitiveness. 'Science for Science' refers to fundamental research, emerging from the scientific community, with 'curiosity' as the presumed driving force. 'Science for Society' implies research that is directed towards solving societal problems, where research questions are defined by societal stakeholders. 'Science for Competitiveness' refers to research originating from industrial agendas to develop new business opportunities, often performed in public-private partnerships. Here too, the stakeholders, together with the researchers, determine the agenda.

However useful such classifications may be to clarify positions, they are in practice a gross simplification of the way scientific research actually works, and therefore not a very useful tool for guiding scientific advancement or indeed promoting useful outcomes. In actual fact, most scientific research is nurtured by multiple sources of inspiration, including curiosity, a longing to understand and explain, as well as the sense of responsibility to address societal challenges. Furthermore, in the Netherlands scientific research is closely linked to (higher) education, adding another source of inspiration.[5]

In the real world of scientific research, these distinctions are, thus, very difficult to make. Take the four winners of the 2015 NWO Spinoza Prizes. These winners were honoured for the excellence of their academic work, because 'according to international standards [they] belong to the absolute top of science. The Spinoza Laureates perform outstanding and ground-breaking research', and 'inspire young researchers'.[6] They are, in other words, first and foremost excellent scholars who, following their curiosity, have managed to produce outstanding work. At the same time, these people do work that is, directly or indirectly, relevant to society. René Janssen's group at Eindhoven Technical University combines physics and chemistry to develop plastic solar cells, which look likely to be produced commercially in the future. Anthropologist Birgit Meyer works in Utrecht on African religions and how they transfer to European contexts in migrant communities. At the University of Amsterdam, mathematician Aad van der Vaart develops statistical models that can help understand the outbreak of epidemics. Cisca Wijmenga, professor of Genetics in Groningen, has

---

5   This has prompted some to propose a fourth category 'Science for Education', referring to the fact that scientific research plays an important role in our education systems.
6   www.nwo.nl/en/research-and-results/programmes/spinoza+prize, consulted on February 21, 2016.

worked on the causes of diabetes, leukaemia, and gluten intolerance. In all four cases, the distinction between fundamental and applied research simply fails to make sense. Paradoxically, this may be especially true at the pinnacle of the research system.

Modern scientific research is predominantly performed in groups and networks, uniting the creativity and inspiration of multiple individuals at different stages in life and from different backgrounds. People interact, discuss options and challenges, leading to mutual inspiration and a stimulating environment which attracts others. Increasingly, these others also include external stakeholders and professionals who may be interested in the research or benefit from it in their own professional environments. In the same way, researchers trained in different scientific disciplines connect and cooperate to combine expertise, for example to address complex scientific and societal challenges.

Such a diverse and dynamic environment responds to a variety of signals, but is always focused on finding opportunities to continue a promising line of investigation. 'Promising' is often defined by a combination of fundamental challenges and opportunities for application, as in the examples of the 2015 Spinoza Prize winners, and the people driving the research combine opportunities to secure the necessary means, making use of various sources of funding. Policy organisations, including funding agencies, often focus exclusively on their direct, individual contribution to the research endeavour, assuming that targeted funding and criteria guide choices by the scientist applying for it. However opportunistic scientists may appear (as any other entrepreneurial individuals), we argue that, in reality, directed funding does not in fact steer much at all. Nor should it. Put in an ecological perspective, it is the seeds growing on the plants themselves that germinate when ready, and grow into diverse blossoms, with nutrition coming from fertile lands. Here, funding is merely a source of water determining its growth rate, but not its cause.

The ecological metaphor implies an important dimension of modern science: interdisciplinary work and, more in general, a denser set of connections between scientific disciplines. For much of the twentieth century, science and scholarship benefited from increased specialization within the clearly defined boundaries of the 'disciplines'. These had common agendas, methodologies, and professional standards, as well as communities of practitioners who were referring to a shared literature. In recent decades, the life sciences in particular have led the way in a process of breaking down barriers and establishing new, interdisciplinary connections and even completely new fields of research – with spectacular successes like the unravelling of the human genome. A similar success story is the emergence of brain and cognitive sciences, where natural and life sciences team up with behavioural sciences. Technological developments are breaking down the barriers between other fields as well. Think of 'big data' with its novel approaches creating new fields that connect but also complement existing disciplines.[7]

The Dutch National Research Agenda: policy or not?

Like science itself, science policy is not static. The NWA is a new string to the bow of research policies. As explained in the introduction of this volume, the Dutch NWA has adopted an unusual approach, and produced an unusual result. It is therefore, in more than one way, an experiment. This has proven confusing for scientists as well as policymakers, perhaps also because of its interactive and uncontrolled process. The NWA was invented and produced on the hoof. Still, we would argue, this experiment blends in very positively with what we consider a healthy research environment.

First and foremost, because its experimental nature reflects in a fundamental way the process of research itself: it has direction, but no premeditated outcome. Its shape as a result defies the type of simplistic classifications often underpinning science policies. Instead of a reductionist approach, the NWA has embraced the diversity of science, the heterogeneity of the Dutch research landscape, and the complexity of the societal challenges the Agenda seeks to address. Its very form, 140 questions, underlines that there are no straightforward solutions to complex problems. This, of course, is also its weakness; the NWA has not produced a list of 'greatest hits' that politicians might embrace. The identification of a limited number of 'routes' has provided such a shortlist, but it would be a pity if the 140 questions that form the body of NWA would be lost from sight as policymakers concentrate on the skeleton of the selected routes.

This brings us to a second asset of the NWA: it is, unlike many science policy measures, inclusive in nature rather than exclusive. It is not primarily about competition, selection, or choices, but instead highlights linkages, synergies, and added value. It aims to connect scientific, societal and economic challenges and brings together scientific disciplines. In doing so, it also straddles the divide between fundamental and applied research.

---

7   European Commission. Validation of the results of the public consultation on Science 2.0: Science in Transition. 2015. Available through https://ec.europa.eu/research/openscience/index.cfm.

In other words, this Agenda harnesses the intrinsic processes in science, rather than forcing it to change course and direction against the grain of scientific practices. We expect that it will therefore be easier to embed in current policies, reducing the transaction costs that are inevitable with any policy reform. It is precisely for this reason, we would like to argue, that the – at first sight unwieldy – list of 140 questions and 27 routes may well prove to be effective. Of course we have to await formal evaluations and can, at this point, not even properly assess its potential. There are, however, promising signs.

This has much to do with what we see as a third strength of the NWA: its potential to connect. In one sense, the NWA has been constructed as a critique of the previous round of major science policies in the Netherlands, the so-called top sectors. This was a top-down initiative that has had mixed reviews, precisely because of commitment problems. Scientists were reluctant to participate in what felt like a coercive collaboration, and for many societal stakeholders it was unclear how they might find partners in the academic community. Especially small and medium-sized businesses, without their own R&D units and science managers, found it difficult to engage. The NWA process has already created inspiring encounters between scientists, policymakers, and professional experts from society and the business community. From these dialogues joint ideas and shared visions emerge on how to move forward. Building on this enthusiasm, we expect new collaborations to emerge that were less likely in the previous, top-down approaches, or indeed in a completely unstructured bottom-up dynamic.

In general, the NWA still falls within the principle outlined in the 1918 Haldane Report in the UK, which argued that the development of science policy, due to its innate complexity, was best left to the experts themselves. This principle has been the foundation of the very successful development of research during the last century.

Conclusion

We have considered some features of the Dutch science system and the merits of sensible science policies. In both, 'diversity' is prominent. The Dutch academic system is strong in many disciplines, and is located in more than a dozen universities, plus another two dozen prominent research institutes. Although some programmes are truly outstanding, there is no evident cluster that can be the foundation for a single-centred policy, in terms of subject or location. A reorientation of policy in that direction seriously risks damaging the whole system, and nipping promising new developments in the bud. The science system can probably improve its societal impact, but not by privileging applied over fundamental research, for the simple reason that cutting-edge research more often than not combines the two. The NWA reflects these features of the Dutch science system and tries to build on them in ways that try to produce new synergies.

References

Dasgupta, Partha, and Paul A. David, 'Toward a new economics of science', *Research Policy* 23, 1994, pp. 487-521

Heinze, Thomas, Philip Shapira, Juan D. Rogers, and Jacqueline M. Senker, 'Organizational and institutional influences on creativity in scientific research', *Research Policy* 38, 2009, pp. 610-23

Koier, Elisabeth, Barend van der Meulen, Edwin Horlings, and Rosalie Belder, *Chinese borden – Financiële stromen en prioriteringsbeleid in het Nederlandse universitaire onderzoek* (The Hague: Rathenau Instituut, 2016)

Laudel, Grit, 'The "quality" myth: Promoting and hindering conditions for acquiring research funds', *Higher Education* 52, 2006, pp. 375-403

Lepori, Benedetto, 'Coordination modes in public funding systems', *Research Policy* 40, 2011, pp. 355-67

Öquist, Benner, *Fostering breakthrough research: A comparative study* (Akademirapport Sweden, Royal Swedish Academy of Sciences, 2012)

Salter, Ammon J., and Ben R. Martin, 'The economic benefits of publicly funded basic research: a critical review', *Research Policy* 30, 2001, pp. 509-32

Volberda, Henk, and Frans van den Bosch, 'Een béétje beter maakt slechter: De rol van sociale en technologische innovatie bij innovatiesucces', *M en O,* 67, 2013, pp. 35-56

Managing what Cannot be Managed

On the Possibility of Science Policy

Barend van der Meulen

Waldsterben. We are familiar with that term mainly from the 1980s, when acid rain led to the death of large tracts of Europe's coniferous forests. But the phenomenon is in fact much older and occurred for the first time in Central and Eastern Europe in the nineteenth century. In the eighteenth century, Frederick the Great of Prussia had developed the concept of the modern state – a state involving itself in all facets of society. This also led to the idea of 'fiscal forestry', i.e. forestry intended to fill the coffers of the state. Pursued systematically, this approach showed that some trees and some areas of forest produced higher yields than others. Birch and spruce, in particular, turned out to quickly yield a lot of timber, leading to the planting of large areas of production forest in Prussia. That was a success, and the approach spread to other parts of Europe. But what the ancient forests had never been able to teach the foresters became apparent in the nineteenth century. The emphasis on rapid production and the resulting monoculture led to the soil becoming exhausted and to *Waldsterben,* i.e. 'forest dieback' – the simultaneous death of large tracts of forest within only a short time.

In his book *Seeing like a State* (1998), James Scott uses the example of fiscal forestry to show that when governments attempt to manage matters, they are doomed to simplification. (Scott, 1998) At first, such simplification appears to offer solely advantages, but in the long run things go wrong. Scott's book – which is subtitled 'How Certain Schemes to Improve the Human Condition Have Failed' – otherwise mainly concerns large-scale modernist projects that we tend to see as being pursued by overambitious technocrats and governments blinded by ideology: Le Corbusier's *ville contemporaine*, the Soviet collectivisation of agriculture, and the enforced creation of villages in African countries. That is a pity. The issue of how government can manage activities about which it actually knows too little is not only important for overambitious technocratic dictatorships. In the Dutch situation too, with a government that is by virtually all criteria normal, it is also relevant, certainly as regards science policy. After all, what does the government know about science?

I know that after reading the above introduction there will be researchers in the Netherlands who will instantly recognize my Prussian forestry example as a metaphor for the country's current system of science and scholarship. This system sees itself compelled by the 'key economic sectors' policy to focus on short-term economic yields (Valorisation! 'From knowledge to skills to cash!') and is slipping into a monoculture because of a lack of appreciation for the humanities and other supposedly economically useless knowledge. I myself don't think things are really all that bad. Despite politicians and science administrators for thirty years now calling for more profiling, choices, 'peaks in the delta', focus areas and all the rest of it, the Netherlands, like other rich Western countries, has a broad portfolio of scientific and scholarly disciplines (Horlings and Van de Besselaar, 2012). The Ministry of Education, Culture and Science is the largest funder of research, and in its relationships with the Netherlands Organisation for Scientific Research (NWO), the Royal Netherlands Academy of Arts and Sciences (KNAW), and the universities it still adheres to the principle of 'managing at a distance'. There is no question of any 'science dieback' in the Netherlands (KNAW, 2015).

The recurring question for the government, however, is how knowledge for specific purposes can be encouraged within the broad palette of Dutch science and scholarship. The 'key economic sectors' are the most recent example of this. At the end of the last century, the aim was to encourage emerging fields such as biotechnology, materials science, microelectronics, nanotechnology, and catalysis. In the 1970s, when the Netherlands briefly had a separate Minister for Science Policy, there was a research policy for such areas as toxicology, soil science, and demography. And ten years before that, scientists had raised the question of how to choose between different investments if an end were to come to the almost exponential increase in the budget for science in the 1950s (Weinberg, 1962,). It was at about that time that some Dutch professors began to push for an explicit science policy, and the former Advisory Council for Science Policy (RAWB) was set up. The initial recommendations by the new council concerned, *inter alia*, participation in CERN, investment in space research, and the possibility of involving persons other than scientists when making choices in the scientific field (National Archive, 2008,).

So there would seem to be nothing new under the sun. But over the years there has been a growing understanding of the relationship between government and science – including the essential tension that there is – and there has been successful science policy. And what seemed successful or promising has come to a stop again. Based on game theory insights into the relationship between government and science and past experience, I shall attempt to analyse what possibilities exist for managing research.

Why have a science policy?

Can government in fact manage scientific research? A legitimate counter-question is whether government should wish to do so. There are good reasons for leaving scientific research in the Netherlands to the free dynamics of science and the choices made by researchers, certainly if the value of knowledge is in itself a sufficient reason to fund research. But governments do not invest in science only in order to *know* – they also want to *win*. And knowledge is a powerful weapon. The US federal budget for research has its origins, *inter alia*, in the American Civil War, the Second World War, and the significantly named 'War on Cancer'. The European research budget received a powerful boost in the 1990s when Japanese companies, with the support of their government, were winning the technological race in microelectronics. European companies such as Philips and Siemens sought and found support for additional investment at European level. And in our own country the ever-present threat of water is still always a good reason to have a world-renowned technological institute in this field, as well as research programmes, and several clusters of questions on the quality, management, and use of water in the Dutch National Research Agenda.

In 1963, the then Minister of Education, Arts and Sciences, Theo Bot, defended the concept of science policy as follows:

> Science has come to occupy a central place in modern Western society, and there is virtually no sector of contemporary society in which it does not exert its influence to a greater or lesser extent. […] At the same time, scientific research demands ever greater financial and human resources. Both these developments have led to science no longer being a matter for scientists in general, but a matter of general interest. […] We cannot do without a deliberate policy. (Handelingen, 1963-64, pp. 729-731)

The Minister went on to note that science policy must also recognize the value of *'free research that is not aimed at increasing the standard of living'*, must create the right conditions for such research, and must ensure that it does not focus too narrowly on the natural and technological sciences. Those are arguments that resonate into the present century.

The Minister also added immediately that

> it [is] still too early to predict the solutions to which the reassessment of the Dutch organisation for science policy, which has now commenced, will ultimately lead. Every country, including the Netherlands, will need

to develop its own institutional and procedural solutions [...] It is not possible to indicate general rules for this [...].

And that brings us back to the core of the problem. Government wishes to pursue policy, and have good reasons for it, but how it should do so is often unclear.

Knowledge for sale

The question regarding 'managing science' can probably best be answered if we start from the simple idea that government must make choices and then ensure that we become the best in the fields it has chosen. And for those who do not wish to leave matters entirely to the government, 'the Netherlands Ltd' should make the choices and we should become the best in those fields. There is a lot that can be said against such an approach. But the idea of making choices as the core of science policy is a constant theme of advisory reports, policy papers, and discussion contributions about science policy. That makes it at least a useful starting point for an essay on the issue of managing science.

To put it in a rather less commercial manner: science is important for the welfare of society and the economy (as was forestry in Prussia). It is therefore an object of state concern, and the state must ensure that Dutch scientists carry out – or are able to carry out – the research that the country needs, and above all the research that the country *really* needs. This creates a relationship between government and scientists that goes beyond the former patronage relationships in which powerful administrators or rich patrons gave artists and scientists the scope to engage in the arts and science.

A factor that should make it easier for government to make choices is that there is a financial relationship between government and researcher. This offers the possibility of a contract between the two parties. However, that contract concerns 'knowledge' and is therefore different to a contract by which government buys a new logo, commissions a new tunnel, or purchases office chairs. Although on the fringes of research policy, parts of applied research become the subject of calls for tenders which seem hardly to differ from the procurement procedures for a logo, tunnel, or office chair, it is generally the case that government knows too little to be able to specify what knowledge it needs. That immediately leads to the first experience-based conclusion: to buy knowledge, you need more than just money.

Buying knowledge requires knowledge of knowledge. In 1972, the British government introduced the 'customer-contractor' principle for funding policy-supporting research. The role of the government as a 'customer' for research needed to be more clearly separated from that of the researcher as a 'contractor' for research projects. In their book *Governance and Research* (1983), Maurice Kogan and Mary Henkel tracked several years of efforts by the Department of Health to implement this principle. The problem was always that in order to determine what its demand for knowledge was, the ministry needed advice from experts – in fact researchers who would ultimately also be the ones to carry out the assignments involved. The book is still instructive with a view to understanding why it is difficult for ministries, including those in the Netherlands, to implement an effective knowledge policy and to manage research.

Science policy as a game

The 'contractual relationship' between government and researchers can be envisioned as a game. The outcome of a game is always dependent on the combination of the players' strategies. In other words, it is not just what the government does that is relevant as regards whether it can manage science, but also how the researchers (or their organisations) respond to government policy. And vice versa. Whether the possibilities and needs to manage science also depend on the research strategies of the scientists. Do they allow the government to manage them? The general consensus among researchers is that they do not. In their view, government policy – at most – explains the problems, but not the capacity to thrive and be successful (Koier et al., 2016). The history of science policy shows, however, that not much new policy is established without the involvement of researchers and scientific organisations. Conditions can therefore be created in which scientists apparently still see the benefits of such policy.

The science policy game displays features of a 'principal-agent relationship', i.e. a relationship between a client (the 'principal') and a contractor (the 'agent') in which the client does not have the knowledge needed in order to know whether it is getting value for money from the contractor. The client can try to control the contractor, but doing so is very expensive. It is much cheaper to trust the contractor. Because the client has insufficient knowledge of the matter, the contractor may also pursue its own objectives rather than committing itself entirely to achieving those of the client.

This is a relationship that frequently applies between a client and a professional. Someone who engages a lawyer trusts that the lawyer will devote his or her efforts to representing the client's case, will actually spend the hours he or she bills on dealing with the case, and will not waste time on a case if it is hopeless. But the lawyer may have his or her own objectives. She or he may in fact consider the case to be hopeless but – needing the fees to keep his or her firm going – will continue to pursue the case even if doing so is contrary to the client's actual interests.

Both players in a principal-agent relationship have two options. The principal can trust or control the agent. The agent may comply with the principal's expectations or pursue its own objectives.

If the game is played endlessly, each of the four combinations of strategies is unstable. In each combination, one of the two players can increase its payoffs by switching strategy. If the government trusts, it is advantageous for the researcher to pursue his or her own curiosity. If the researcher pursues his or her own curiosity, it is more advantageous for the government to build in control mechanisms. If the government controls strictly, the researcher better comply with the requirements. If the researcher complies with the requirements, it is more advantageous for the government to trust... (Van der Meulen, 1998). In terms of game theory: a Nash equilibrium is lacking.

Where social relationships are unstable but considered necessary, institutions arise that make possible a long-term relationship (North, 1990). One of these is the idea expressed by Vannevar Bush in his much-quoted report *Science, The Endless Frontier* (1945). Bush wrote the report during the Second World War, which in America led to a number of major mission-oriented research projects intended to contribute to victory in the short term. The most familiar example is the Manhattan Project, which produced the atom bomb. The reason for Bush's report was President Roosevelt's question as to whether – once the military struggle had ceased – the R&D expenditure involved should be continued in peacetime, and if so, how. The core message of the report was that government must continue research efforts in peacetime too, but based on the trust that basic research can push back the boundaries of knowledge and will be useful in the long run. That trust could not be blind, however: to distribute funds a National Science Foundation had to be set up, headed by scientists who would ensure that the money found its way to the best researchers.

That was not an obvious matter. It took a long time before Congress approved the idea that not government itself would decide on the distribution of the money but rather those who received it. But for a government that actually dares to trust that scientific research will generate yields in the long term, this is a favourable configuration of players and strategies. Thanks to competition and mutual quality control between researchers, the government can be confident that the money will find its way to the right place and will pay off in the long run. For researchers, this configuration provides the scope to pursue their own goals and is therefore interesting. As long as the cost of the 'control' – i.e. writing and evaluating proposals – remains within bounds and enough researchers are allocated funds, there is a Nash equilibrium. In the Netherlands, this combination of players and strategies has evolved into what we now know as the NWO's Open Competition, and at European level into the European Research Council (ERC). However, the pressure on the contract has become extremely high. Even researchers who are clearly benefiting, or have benefited, wonder whether 'the system' has not gone too far and whether the perverse effects are not overwhelming the benefits (Science in Transition, 2013; Bollen and Scheffer, 2015).

Development of the 'first flow of funds' in the Netherlands (i.e. direct government funding) shows a similar dynamic of institutionalization of confidence in evaluation practices. For a long time, financing of university research was based on blind trust. Funds were provided but there was no monitoring of whether they were utilized effectively. In 1979, the Minister pointed out in the Policy Memorandum on University Research [*Beleidsnota Universitair Onderzoek*] that it was unclear whether research funds were being spent efficiently: there was no division of tasks, the choice of themes for research was not sufficiently determined by social needs, and it was unclear whether performance was sufficient. What followed was a decade of division of tasks, concentration, selective contraction and growth, advisory committees and conditional funding, and a great deal of frustration in the relationship between government and university researchers. But it did also lead to quality control at universities – in the form of external reviews of research, output indicators, and priorities – becoming something commonplace (Whitley and Gläser, 2007). The government's trust is no longer blind but is well-founded. Some 35 years later, the Minister could inform the Dutch House of Representatives that *'the Dutch scientific system performs extremely well in terms of quality and productivity'* (Minister of Education, Culture and Science, 2015).

Aiming for socially relevant goals

For situations in which government does have its own goals, this configuration of well-founded trust is not a satisfactory solution. Neither through competitions organised by research councils such as the NWO, NSF, and ERC nor through control mechanisms such as external reviews of research and output indicators can government manage matters in such a way that 'the right kind' of research takes place, in other words, research that 'meets the needs of society', or, as it is termed in the recent *Vision for Science 2025* document [*Wetenschapsvisie 2025*], 'research with maximum impact'.

One of the ways of achieving this is to organise competition oneself. In the UK, the government does this by means of the Research Excellence Framework. The remarkable thing about the Framework is that, although assessment is carried out by researchers, the criteria, and therefore what is considered to be high-quality research of the right kind, are formulated by the government. The REF describes in detail how research will be assessed, what criteria will apply, and how the universities can demonstrate that their research meets those criteria. To assess the impact of research, for example, a precise definition is provided, a five-point scale, and guidelines for case studies that can demonstrate the impact (HEFCE, 2011).

From the perspective of game theory, there was a similar situation in the Netherlands after 1994 in the distribution of money from the Economic Structure Enhancing Fund (FES) (a fund created from natural gas revenues) which was destined for the knowledge infrastructure. The 'Committee of Sages' which advised how the money should be spent, assessed the proposals for research programmes not only according to the quality of the research involved but also according to such things as whether the programmes displayed cohesion, whether the intellectual property, demand-driven nature, and anchoring of the results were properly arranged, whether management was effective, and whether there was international entrenchment. The committee continued to keep track of the programmes while they were running, and did not hesitate to issue recommendations for improvement on the basis of interim reports (Commissie van Wijzen Kennis en Innovatie, 2011).

Past experience shows that it is difficult to ensure that such a configuration remains stable. In the UK, every new round of evaluation leads to heated discussion and changes in policy. In the Netherlands, researchers complained that distribution of the FES funds was politically determined. When a new cabinet was formed in 2010, the government terminated the FES programmes so quickly and in such an ill-considered manner that the realisation is only now beginning to dawn that it was thus surrendering one of its last methods for managing research.

Agendas as an instrument for control

A third way to arrive at a workable relationship between government and researchers is to reach consensus on the priorities for research. If government and researchers agree on the goals, government can minimize the cost of monitoring whether the researcher is doing the right research, while the researcher can still investigate what he or she is curious about. This way of organising the relationship between government and researchers has a long tradition in the Netherlands. Because the Dutch National Research Agenda is in line with that tradition, it is a good thing to once more reconsider this historical line in science policy.

Agenda setting in Dutch science policy began in agriculture. The National Council for Agricultural Research (NRLO) was established in 1957 as part of the Netherlands Organisation for Applied Scientific Research (TNO) in order to coordinate agricultural research. A compromise organisation, the TNO legislation did provide for a TNO agriculture organisation but the Ministry of Agriculture did not want its institutes to be made part of TNO. The NRLO therefore became a council within TNO that had to coordinate research outside TNO, a seemingly impossible construction. Nevertheless, the NRLO continued to exist for a remarkably long time and was only abolished in 2000. By then it had ceased to be a council within TNO and had evolved into a combination of a sector council and an agency of the Ministry of Agriculture, Nature and Food Quality (Dijksterhuis and Van der Meulen, 2007).

The history of the NRLO also shows how difficult it is to manage research. Initially, the Council tried to do so with typical planning instruments. It categorized the research, mapped out who did what, set up a project administration, and developed criteria for the economic evaluation of the research. This did not lead to the expected rationalisation which would allow priorities to be set on the basis of clear cost-benefit analyses. The development of 'social frames of reference', also taking account of 'psychological' and 'legal aspects', did not help either. In the course of the 1970s, the rationalistic approach slowly disappeared into the background. That approach led to a great deal of paperwork regarding tasks which – as the staff noted – could also be performed in a qualitative manner by experts and other persons involved.

That this shift could happen was also because the NRLO had gradually evolved into a participation organisation. Its field of activity had been extended to the whole of agricultural research, including research taking place within, for example, Wageningen Agricultural University and the Faculty of Veterinary Medicine in Utrecht. 'The sector' also became increasingly involved in the consultations through the then Chamber of Agriculture and the Product Boards. In the 1970s, environmental organisations and other ministries were also given seats on the Council. The NRLO evolved into an organisation made up of larger and smaller consultative bodies within which different interests and perspectives regarding agricultural research and innovation were brought together, weighed up, and translated into recommendations for priorities and research programmes.

The 1974 Science Budget presented this approach as exemplary for all sectors of Dutch society. Such a comprehensive system has never been implemented. Development of the system of sector councils was fraught with difficulty and it was only in 1986 that the government submitted the Sector Councils for Research and Development Act [*Wet Sectorraden Onderzoek en Ontwikkeling*] to Parliament. Besides the NRLO, only three sector councils gained a firm foothold: the RAWOO (research in the context of development cooperation), the RMNO (environment and nature), and the RGO (health research). Other sector councils were fairly quickly disbanded, for example the Council for Energy Research and the VRA-OGO (research on the built environment), or got stuck at the concept stage (Council for Technological Industrial Research, Sector Council for Chemical Research). By 2000, the sector councils were among the first to perish in the drive to reduce the number of bodies advising government. That they had been extremely valuable in managing research played no role in the decision-making on this matter.

The unique thing about the sector councils was that, very early on in the history of science policy, they institutionalised the idea that consultation would enable government, research organisations, and civil-society organisations to develop research objectives and thus agree on the management of research. This approach produced a range of instruments that deployed long-range strategies providing a political explanation of how the research fitted in with government policy objectives and the needs of society, and vice versa how government and the sector concerned could respond to developments in research. This in turn led to a series of national research programmes in the areas of agriculture, the environment and nature, and healthcare.

The same approach was also adopted in the 1990s by the Consultative Committee on Foresight Studies [*Overleg Commissie Verkenningen*], which was instructed by the then Minister, Jo Ritzen, to organise foresight studies and identify priorities for scientific research (OCV, 1997). The activities and reports of the committee covered a wide range of disciplines and research areas, including chemistry, art history, economy, and nanotechnology. After four years, following its final report, the committee was disbanded. One of the experiences of the consultative committee was that the blessings of science policy were not confined to the sciences and engineering fields. Among its more successful initiatives was the panel on the future of research in law. The panel observed that current research in law did not match with the needs of stakeholders and major societal developments such as globalisation, high tech innovations and deregulation. As a result, the government put law research high on the agenda in its Science Budget 1997; the research council was asked to set up research programmes to support new research programmes on law and funding became available for a research institute on the internationalization of law.

More recently, a number of the FES programmes, such as the Climate Change Spatial Planning Programme [*Klimaat voor Ruimte*] and Knowledge for Climate [*Kennis voor Klimaat*] generated methods for programming and also implementing research together with civil-society stakeholders. This institutional entrenchment of the approach disappeared, however, along with the FES programmes. In the context of innovation policy, 'tripartite consultation' has been modernised as the 'Golden Triangle', and within the key economic sectors long-range strategies and research agendas are referred to as 'roadmaps'. The idea of joint consultation remains, although the government seems to have hardly any innovation objectives of its own, and has placed implementation and the steering wheel in the hands of industry.

In conclusion

What does this analytical and historical consideration of the management of science tell us about current science policy? Firstly, that government does wish to pursue science policy. However, the legitimacy of the idea of science policy is based on the importance of science itself rather than on a clear perception of what government can do to manage the system.

Researchers too want government to pursue a science policy. One of their main reasons is that science always seems to grow faster than the government's budget. That applied to the 'big science' of large-scale facilities (De Solla Price, 1963), but nowadays it also applies to the 'little science' of individual applications for VENI scholarships (Van Arensbergen et al., 2013) and art history (KNAW Verkenningscommissie Kunstgeschiedenis, 2013). If researchers experience bottlenecks in their discipline, career possibilities or research infrastructure, advisory reports quickly call upon government to help. In such cases even scientists apparently have greater trust in government than in their own organisations and colleagues.

For many years, two tracks were visible in science policy. One is today referred to as 'excellent science', but it began, once upon a time, with the despairing conclusion by the Minister of Education, Culture and Science that it was unclear just what the quality of scientific research in the Netherlands actually was. Measures aimed at improving quality can almost always count on support from the scientific community and – almost as a matter of course – from the scientific elite. Thanks to the NWO's Open Competition, Innovational Research Incentive, Spinoza Prizes, review committees, and 'Gravitation' schemes, government's trust in the quality of research has been firmly institutionalised. This part of science policy provides a simple answer to the initial question whether government can manage science. Yes it can, by institutionalising proper evaluation and funding processes run by scientists and their organisations. Scientists are said to be curiosity-driven. One of the lessons of science policy is that monitoring performance and quality assessments are helpful as well to drive researchers.

The second track is a winding and bumpy one, and one that seems to constantly find itself at a dead end: the targeted encouragement of research areas. This track shows how unstable the relationship is between government and science. Theoretically, institutions can be designed, such as competition and consultation, to make such a policy possible. Looking back, we see that these have in fact been used, in the form of advisory committees, committees of sages, sector councils, and foresight studies, but none of these instruments has become permanently institutionalised. In the Dutch system, it is currently the European Commission that is perhaps the only authority which manages research towards specified aims. A significant amount of the money in the Horizon 2020 research programme is divided on the basis of Grand Challenges, but it is questionable whether that will remain so. European research policy too has from the outset sought an effective way of managing research.

This may sound somewhat defeatist, but it may in fact be the most important lesson for the Dutch Research Agenda. History shows us that the emergence of new fields of research, the urge to push back the boundaries with larger facilities, and society's need for relevant knowledge have always been a reason to develop ad hoc solutions to what is a persistent problem. For a long time, the Netherlands had various consultation structures – between government departments, between government and researchers, between researchers, government, and civil-society parties – from which solutions were slowly but surely developed in the form of research agendas, long-range strategies, and research programmes. There is a 'Vision for Science' document [*Wetenschapsvisie*]. There is a research agenda. Let us hope that from these, research programmes will evolve in which government, civil-society organisations, and researchers will be able to unite their various interests.

Bibliography

Arensbergen, Pleun van, Laurens Hessels, and Barend van der Meulen, *Talent Centraal. Ontwikkeling en selectie van wetenschappers in Nederland* (The Hague: Rathenau Instituut, 2013)

Bollen, Johan, and Marten Scheffer, *De wijze massa, waarom wetenschappers onderzoeksgeld beter zelf kunnen verdelen,* De Omslag, 2015, https://omslag.nu/onderzoeksfinanciering/de-wijze-massa/ (retrieved 5 March 2016)

Bush, Vannevar, *Science, The Endless Frontier* (Washington: Office of Scientific Research and Development, 1945)

Commissie van Wijzen Kennis en Innovatie, *Advies aan het Kabinet over de resultaten van de totale BSIK-impuls,* 2011

De Solla Price, Derek, *Little Science, Big Science... and Beyond* (New York: Columbia University Press, 1963)

Dijksterhuis, Fokko Jan, and Barend van der Meulen, *Landbouwinnovatie en onderzoekcoördinatie, De Nationale Raad voor Landbouwkundig Onderzoek, 1957-2000* (Groningen/Wageningen: Netherlands Agricultural Historical Institute [NAHI], 2007)

Handelingen der Staten Generaal 1963-1964, II, pp. 729-731, www.statengeneraaldigitaal.nl

HEFCE, *Assessment framework and guidance on submissions,* REF 02.2011, 2011

Horlings, Edwin and Peter van de Besselaar, *Focus en massa in het wetenschappelijk onderzoek: de Nederlandse onderzoeksportfolio* (The Hague: Rathenau Instituut, 2012)

KNAW Verkenningscommissie Kunstgeschiedenis, *Verschilzicht. Beweging in het kunsthistorisch onderzoek in Nederland* (Amsterdam: KNAW, 2013)

KNAW – Adviescommissie Witte Vlekken Universitair Onderzoek, *Ruimte voor Ongebonden Onderzoek, signalen uit de wetenschap* (Amsterdam: KNAW, 2015)

Kogan, Maurice and Mary Henkel, *Government and Research. The Rothschild Experiment in a Government Department* (London: Heinemann Educational Books, 1983)

Koier, Elisabeth, Barend van der Meulen, Edwin Horlings, and Rosalie Belder, *Chinese borden – Financiële stromen en prioriteringsbeleid in het Nederlandse universitaire onderzoek* (The Hague: Rathenau Instituut, 2016)

Meulen, Barend van der, 'Science policies as principal-agent games. Institutionalization and path dependency in the relation between government and science', *Research Policy* 27, 1998, pp. 397-414

Minister of Education, Culture and Science, *Wetenschapsvisie 2025, Keuzes voor de Toekomst* (The Hague: Ministry of Education, Culture and Science, 2015)

National Archive The Hague, *Inventaris van het archief van de Raad van Advies voor het Wetenschapsbeleid (RAWB)*, 1966-1990, No. 2.14.82, 2008

North, Douglas, *Institutions, Institutional Change and Economic Performance* (Cambridge: Cambridge University Press, 1990)

Overleg Commissie Verkenningen (OCV), *Een vitaal kennissysteem, Nederlands onderzoek in toekomstig perspectief* (Amsterdam, 1996)

Science in Transition, *Waarom de wetenschap niet werkt zoals het moet, en wat daar aan te doen is,* Position Paper, 2013, www.scienceintransition.nl/wp-content/uploads/2013/09/POSITION-PAPER-16-sep-2013.pdf (retrieved 5 March 2016)

Scott, James, *Seeing Like a State, How Certain Schemes to Improve the Human Condition Have Failed* (New Haven: Yale University Press, 1998)

Weinberg, Alvin, 'Criteria for Scientific Choice', *Minerva* 1 (2), 1962, pp. 158-171

Whitley, Richard, and Jochen Gläser, The Changing Governance of the Sciences, *Sociology of the Sciences Yearbook 26,* 2007, Springer

The Art of Making Connections

Ed Brinksma

Introduction

In this chapter, I consider the questions surrounding the Dutch National Research Agenda more specifically from the perspective of universities of technology. To what extent does their traditionally closer alliance with industry and their track record in the application of science imply different or more specific policies regarding the programming and management of their research portfolios? Compared to comprehensive universities, universities of technology tend to put a relatively greater emphasis on their responsibility to make their research have societal and, more specifically, economic impact. The universities of technology in the Netherlands have formalized this by recognizing this objective, referred to in Dutch as *valorisatie*, as the third main ingredient of their mission after education and research.

Given the relevance of applications and impact it is often thought that a fundamental approach to science is of less relevance for the research portfolios of universities of technology, and the role of free and independent research is consequently smaller. I will argue that these are misconceptions and that, quite on the contrary, basic research plays a vital role for universities of technology. Their more specific mission, however, does imply a special position and role in connecting science, technology, and society.

This is an interesting observation in the context of one of the more surprising outcomes of the Dutch National Research Agenda, namely the connections, or 'routes', that have been identified as productive and interesting links between different research questions. In fact, the current trend towards more integral research programmes, enabled by old and new multi-disciplinary connections, will create a new scientific ecosystem. In this system some of the particularities of research policies for universities of technology seem to carry over to other parts of the system as well, especially where basic research and applications have meaningful encounters. I will argue that the Dutch National Research Agenda with its catalogue of the 140 leading research questions provides just one of the many dimensions in which such connections must be made.

The impact of World War II: the linear model

It is not our intention to give a full historical account of the policies regarding academic research and its application, but it will be useful to revisit the main ideas that developed in the period that has been most relevant for the state of affairs of today, namely the time since World War II, including the war period itself. In the Second World War the United States established itself as the undisputed leader of the West in many respects, including national research policy. A huge, influencing factor was, of course, provided by the unprecedented contribution of research to the war effort, in part fuelled by the fear that Axis powers would be the first to develop an atomic bomb. In 1941 the Office of Scientific Research and Development (OSRD) was established by Executive Order, and given almost unlimited access to funding and resources. In addition to the highly classified Manhattan Project, which developed the first atomic bombs, the OSDR oversaw a wide variety of projects, including work on guided missiles, radar, early-warning systems, and more effective medical treatments.

A central figure in these developments, and what followed later, was OSDR director Vannevar Bush (Zachary, 1997). Vannevar Bush was an electrical engineer who worked on the first analog computers for solving differential equations and served as vice-president of MIT and dean of the MIT School of Engineering. One of his notable students was Claude Shannon, the father of information theory. Bush, however, would achieve his greatest renown as an extremely effective science administrator. As director and prime instigator of the OSDR, Bush saw to it that he reported directly to the President, and as such was in effect the first presidential science advisor.

Towards the end of the war, in the summer of 1945, Bush tried to capitalize on the enormous prestige that he built up during the war effort through his report to the president, *Science, The Endless Frontier* (Bush, 1945). Its purpose was to create a successor to the OSDR that would secure substantial funding for research in peacetime (Greenberg, 2001). In particular, he proposed that basic research is 'the pacemaker of technological progress', promoting the view that basic research is the principal source of technological innovation. This view on the dynamics of technological innovation later became known in a more extended version as the so-called *linear model*, which is depicted in Figure 1.

The linear model asserts that innovation advances by a dynamic flow from science to technology has been very influential, and has been used as a guiding principle of R&D managers the world over. The National Science Foundation, which can be seen as the successor to the OSDR that Bush

Figure 1   The linear model

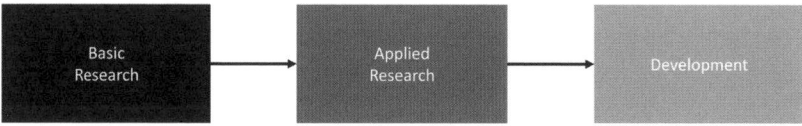

advocated, described this process as the *'technological sequence'* in an early publication (National Science Foundation, 1952, pp. 11-12). It states that this sequence consists of the following three stages:
basic research: 'Basic research, directed simply towards a more complete understanding of nature and its laws, embarks upon the unknown [...]';
applied research: 'Applied research concerns itself with the elaboration and application of the known'; and finally
development: 'Development [...] is the systematic adaptation of research findings into useful materials, devices, systems, methods, and processes [...]'.

It then points out that there is an obvious dependency of the successive stages on the preceding ones. Although the linear model, as we shall see, is an incomplete and inaccurate account of the dynamics of science and technology, it remains until today a very influential conceptual model for technical innovation by virtue of its simplicity. Basic research as the principal source of technological innovation, and the categorization of research in basic, or fundamental, and applied can still be found in many accounts of the innovation process, whether by administrators or by scientists themselves.

Adding another dimension: Stokes' quadrants

Although there have been a good number of earlier critics of the linear model, it was Donald Stokes who proposed a substantial revision of the account of technological innovation in his book *Pasteur's Quadrant – Basic Science and Technological Innovation* (1997). Stokes, a political scientist and dean of the Woodrow Wilson School at Princeton, was bothered by the linear model's strict separation between basic and applied science, which he considered paradoxical: 'The annals of research so often record scientific advances simultaneously driven by the quest for understanding and considerations of use that one is increasingly led to ask how it came to be so widely believed that these goals are in tension and that the categories of basic and applied science are radically separate.' Stokes' prime example of a researcher simultaneously motivated by fundamental curiosity and applicative need is the French microbiologist Louis Pasteur.

I would like to add that the whole concept of innovation as proceeding from basic research is unhistorical, with many prominent counterexamples. The invention of the steam engine by Thomas Newcomen and John Cally in 1705, and its refinement by James Watt in 1724, for example, was a product of empirical engineering, whose scientific underpinning by a theory of thermodynamics was achieved only much later, starting with the work of Sadie Carnot in 1825. Such examples are not restricted to older parts of engineering. In his book *What Engineers Know and How They Know It* (1990) Stanford aeronautical and aerospace engineering professor Walter Vincenti gives a number of examples of problems in aerospace engineering not addressed by any natural science, e.g. control-volume analysis and propeller design and selection. In his *Sciences of the Artificial* (1969), Nobel laureate and Turing Award winner Herbert Simon even argues that the design and engineering of artefacts has its own separate methodology and empirical foundation.

Stokes (1997) observed that it is the one-dimensionality of the linear model that makes it unavoidable that moving towards applied research implies moving away from basic research and vice versa. He proposes to overcome this predicament by moving to a two-dimensional space of possibilities: one dimension representing the extent to which research is motivated by a need for fundamental understanding, and the other representing utility as the driver of research. If each dimension is measured using a simple high/low scale, this creates the space consisting of four quadrants represented in Figure 2.

Stokes labelled three of the four quadrants with the names of well-known researchers/innovators who are representative of the type of research covered by the corresponding quadrant:

Bohr's quadrant: this is the area of pure, basic research, i.e. motivated by a quest for fundamental understanding, and not by considerations of use, as arguably the drivers for Bohr's work on quantum mechanics.

Edison's quadrant: this is the complementary area of pure applied research, i.e. motivated by the need to solve practical problems, and not having any pretence in providing fundamental understanding of the phenomena at hand. This, of course, seems an appropriate characterization of the work by Edison on the development of electrical utilities.

Pasteur's quadrant: this is the new type of research allowed for in Stokes' approach, viz. research motivated by both fundamental and applied objectives, and, as already indicated, he saw the work of Louis Pasteur in microbiology and medicine as a fine example of this kind of work.

Figure 2 Stokes' quadrants

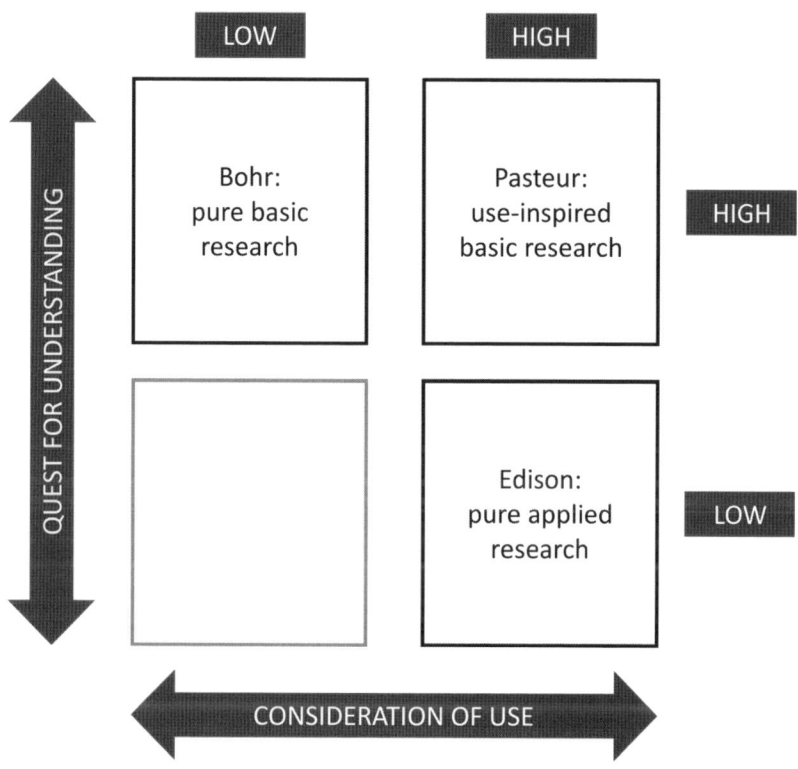

Although Stokes does not dwell on the interpretation of the fourth, unlabelled quadrant, which seems unattractive for research, being motivated neither by fundamental nor applied motives, it is not without scientific significance. It could be used to represent activities such as systematic classification and structuring in the early stages of a new field of scientific inquiry. Some label it *Linnaeus'* quadrant, named after the founder of modern taxonomy, the physician, botanist and zoologist Carl Linnaeus, whose work laid the foundations, among other things, for the work of Darwin on the paradigmatic changes regarding the evolution of species.

Stokes' quadrants offer a better vocabulary for the description of research in the innovation chain than the linear model. Roughly speaking, the research portfolios of traditional universities are dominated by activities in Bohr's quadrant, those of universities of technology by Pasteur's quadrant, and industrial R&D by Edison's. It should be stressed, however, that this is not an exclusively one-to-one relationship. We will see, in fact, that it is quite essential that this is not the case.

Although the *aficionados* agree that Stokes' quadrants give a better account of types of research that exist, it cannot be said that it has effectively replaced the terminology of research policymakers at large, and there remains a tendency to use the terminology of the linear model, probably by virtue of its simplicity. And in the case of exceptions, Pasteur-like research is sometimes misread as a way to boost both fundamental and applied research for the same money (Zijlstra, 2011). As one might suspect, it is not as simple as that.

Dynamics of science and technology

The linear model not only categorizes its research stages, but also proposes a model of interaction between them, namely a transfer from basic research, through applied research to development. Likewise, Stokes proposes a model that describes the interaction between the quadrants, and with that a much richer account of the interaction between science and technology. It is depicted in Figure 3.

In his explanation of this model Stokes quotes Harvey Brooks' observation that 'the relation between science and technology is better thought of in terms of two parallel streams of cumulative knowledge, which have many interdependencies and cross-relations' (Brooks, 1994, p. 479). The parallel, cumulating streams are those of scientific understanding on the one hand, and those of technological development on the other. Stokes sees Pasteur's category of use-inspired basic research as an important link between these streams, taking its cue from both existing scientific knowledge and technology, and making contributions towards the improvement of both.

The great advantage of this model is that it accounts much better for the historical and actual interaction between the different kinds of research and technology than the unidirectional linear model. Watt's steam engine as a technology that inspired Carnot to formulate a theory on the hypothetical efficiency of such machines, which in turn could be used to improve the technology, and at the same time gave rise to the full-blown theory of thermodynamics as we know it today. Or starting at the other end, the quantum mechanical theory of electrons that suggested the possibility of

Figure 3 Stokes' dynamic model

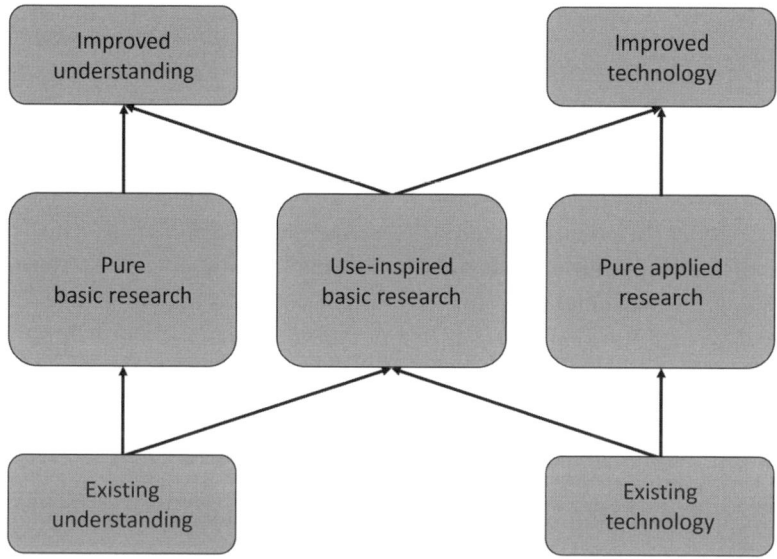

electron microscopy, which once it had become an established technology has made huge contributions to the growth of science itself.

One of the important consequences of Stokes' dynamic model is that it shows the necessity to invest in all types of research (Bohr, Pasteur, and Edison) to obtain successful chains of innovation. Among others things, this yields a much more credible defence for investing in basic research than the historically incorrect claim by Bush and others that all innovation derives from basic research. Conversely, it also shows that policymakers cannot get away by just betting on one type of research alone. In particular, those concerned with the economic proceeds of research and development often advocate concentration on Edison-, or as quoted above, Pasteur-like research. This ultimately leads to suboptimal results for lack of new scientific inspiration. The latter point was what the linear model was 'designed' to explain, a point that is retained in Stokes' approach. Interestingly enough, Stokes' model also explains that a disconnect between science and technology would be to the detriment of science itself, even if basic research were adequately funded, because of a less efficient feedback of technological improvement into the scientific process itself.

In the context of the Dutch National Research Agenda the following quote seems in order here: 'A clearer understanding by the scientific and policy communities of the role of use-inspired basic research can help renew the compact between science and government, a compact that must also provide support for pure basic research' (Stokes, 1997, p. 89).

The role of social sciences and humanities

The compact between science and government that was envisaged by Vannevar Bush did not include social sciences and the humanities; he saw only the exact sciences and medicine as the drivers of innovation (Zachary, 1997, pp. 91-95). Interestingly enough, Donald Stokes, a social scientist himself, does not give much attention to this restriction in Bush's thinking, and concentrates in his criticism of Bush's model on the interaction between science and technology, not seizing upon the opportunity to look at the wider societal context.

There is a growing awareness, however, that social sciences and humanities have a substantial role to play in the process of innovation, and more specifically also in technical innovation. There is a growing list of contextual concerns that are essential for innovations to work, such as ethical, legal, economical, organizational, and psychological aspects, to name just a few. It is from this perspective that social sciences and humanities have been included in *Horizon 2020*, the research agenda of the European Union (European Commission, 2011).

It would seem that knowledge of how to make things work from those contextual perspectives can be regarded as some sort of 'social technology', which has a similar relation to, and interaction with, basic research in social sciences and the humanities as technology has to basic science in Stokes' model. Figure 4 depicts a proposal by the author for an extension of that model that incorporates this interaction.

This extended model implies that social sciences and humanities are part of the interaction governing the progress of the innovation process, and should be taken into account as such. As a matter of completeness, it is good to point out that also in non-technological contexts it makes sense to consider Bohr-, Pasteur-, and Edison-like research types for the social sciences, with interactions between basic research and the design and implementation of interventions.

Figure 4 Extension of the Stokes model

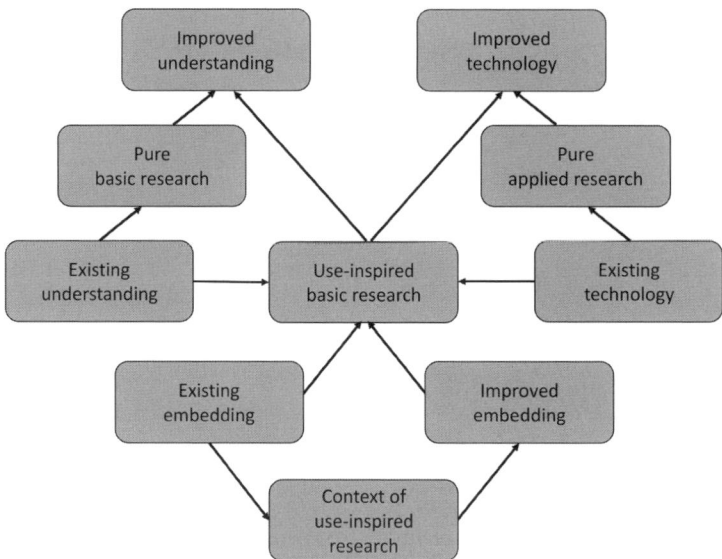

Making connections

The character of universities of technology naturally entails that a substantial part their research portfolio is informed by a variety of application contexts, whose selection depends on the more specific profile of the institution. As already mentioned, universities of technology will, as a rule, concentrate on fundamental research in the applied context, i.e. research in Pasteur's quadrant, developing generalizable knowledge on the relevant application areas, as opposed to the more pragmatic and short-term Edison-type R&D that is typically carried out by non-academic institutions and companies.

In this context Stokes' dynamic model has direct consequences for the development of a sound research policy. It implies the strategic importance of productive links of research portfolios of use-inspired basic research with related Bohr-type basic research on the one hand, and with Edison-type applied research and development on the other. The latter connection is usually available through collaboration partners in the various application domains, where industrial R&D labs are a traditional case in point. The vital connection to Bohr-type research must be ensured by also investing in relevant pure basic research, both directly, as part of own research programmes, and indirectly, through collaboration with other academic partners. As a consequence of this line of reasoning, we witness a growing number of institutional consortia between academic knowledge institutes built around such and other complementarities. As pointed out earlier, the growing significance of the social-scientific context in the success of innovations also warrants investment in creating viable connections with research in social sciences and humanities. With research policies that create effective interactive connections both between the different types of research and different disciplines universities of technology play a pivotal role in the innovation chains of society.

Curiosity versus application

One issue that often appears in the context of use-inspired research, whether basic or purely applied, is to which extent it restricts the freedom that is associated with high-quality research. Such freedom is often seen as *a conditio sine qua non* for curiosity-driven, excellent research. Vannevar Bush's statement that 'applied research invariably drives out pure' is certainly suggestive in this context (Bush, 1945, p. 83). It is also apparent in a certain *l'art pour l'art* (and often romantic) type of defence of pure research, such as in Hardy's *A Mathematician's Apology* (1940), which even suggests that applied mathematics would be 'boring' (Hardy was, of course, blissfully unaware of the future relevance of his own 'pure' field of number theory in modern applications in cryptography). Stokes' example of the research of Louis Pasteur is a very convincing counterexample to this suggestion. A more general argument against it is that it seems to confuse individual motives for doing research with programmatic ones. Good research is always driven by the curiosity of the researcher regardless of its quadrant type. And researchers engaged in basic research most often find themselves embedded in programmes whose objectives have been defined by others. The accommodation of the freedom and creativity of the individual researcher, therefore, is an important aspect of the organization and implementation of research projects, whether basic, use-inspired, or both. A fundamental researcher like Alan Turing did excellent work as part of the war effort in the *Enigma Project* in the UK (Hodges, 1983). The debate over whether working in a particular context of application generally hinders or inspires high-quality research, is as moot as deciding whether blank verse generally makes for better poetry than, say, sonnets.

On the higher aggregation levels of research programming, such as with the setting of research agendas, freedom and quality are linked in another way. The insistence of the research community that regardless of the choices for a National Research Agenda there should be room for free and independent research should not be defended in terms of individual freedom, but rather as a necessary optimization strategy in the face of the inherent uncertainties in long-term research programmes. It can provide alternatives where programmed research gets stuck and fails to deliver. In stochastic optimization theory it is well-known that, in the context of incomplete information, search strategies that involve some degree of random variation generally yield better results than purely deterministic strategies. Or, to give a more concrete intuition: if you do not know which gambling machine in a casino has the highest payout, it is better to spend your money on several of them than to play just one machine, the so-called *multi-armed bandit problem* (Robbins, 1952). Independent research provides the necessary variation in the context of programmed research, and is therefore a natural and necessary counterpart of programmed research.

Another connection that is relevant in this context is the difference between *evolutionary* and *radical* design, as explained in the book by Walter Vincenti mentioned earlier (Vincenti, 1990). Evolutionary design works with the steady improvement of existing design solutions, whereas radical design works with disruptive improvements that involve paradigmatic changes that are typically not associated with strongly programmed research, as this has a natural predisposition for evolutionary approaches. They are more often than not the result of fundamental or free research activities. The notions of evolutionary and radical design go back to the fundamental ideas by Joseph Schumpeter, who introduced the concept of innovation in his seminal work *Theorie der Wirtschaftlichen Entwicklung* (1911). Schumpeter argues that economic development often occurs in shocks, instead of gradually, and that such shocks can often be attributed to completely new insights and knowledge.

Institutional profiles

University research portfolios are usually the result of strategic agendas that refine the university profile into thematic areas considered relevant for a given period of time. These agendas are filled by bottom-up research programming building on existing research strengths and opportunities.

There is a mutual dependency between the profile, the themes, and the actual research done: strategic choices can strengthen new themes and projects, and success (or failure) in research can lead to adaptation of strategy and profile. In the not so distant past there was less emphasis on strategic profiling for universities. Although comprehensive universities and universities of technology were seen to be clearly different, there was a general inclination to consider institutions within the same category as more or less equivalent. More recently, there has been a growing awareness that universities should also develop a more articulated profile, as advocated in Veerman (2010). Some of the drivers for this change in attitude are:

Exponential growth of content: the fast growth of academic knowledge and educational programmes has made it impossible to keep up the promise of a traditional *universitas studiorum* providing access to all scientific disciplines, and therefore coherent choices must be made. For universities it is essential that their research portfolio is consistent with the requirements of their educational programmes.

Globalization: the rapid globalization of higher education makes it important to emphasize the institutional *added value* of the world wide web of universities. Without a distinguishing institutional profile, even universities with good research and educational programmes can become redundant.

Resources and infrastructure: limitation of resources is a common driver of institutional profiling, promoting the selection of those profiles that are most competitive within the available means. Conversely, access to, and investment in, a competitive research infrastructure is a powerful instrument for maintaining and strengthening the profile.

There is a tendency to try and optimize the landscape of university profiles for efficiency, reducing overlap and fragmentation by institutional specialization and concentration. A first priority, however, should be a sufficiently rich and robust system. Of course, excessive overlap and fragmentation can be dysfunctional, but a certain redundancy and variation in scale can be instrumental in increasing the scope and robustness of academic research.

Universities of technology have a clearly distinct profile among universities, both in terms of their orientation on technical sciences and technology, and in terms of their preference for Pasteur-like research programmes. Because of this, as already pointed out above, their relevance is not restricted to the contributions in the technological domains of science and society, but also as providers of strategic connections in both scientific and innovation processes.

Governments, universities, and industry

So far we have looked at research and innovation in the context of the dynamics of different sorts or research and universities. But, of course, this is only a part of the picture. The interaction between knowledge institutes and other societal actors, such as governments and industry, has great influence on research programming and innovation processes.

What do universities, especially those engaged in application-driven research, look for in the agenda-setting activities of their governments, whether at the regional, national, or European level? Of course, governments at any level define their societal priorities, which in turn can be linked to application areas for research. These areas, in turn, can give rise to research programmes involving any or all of Stokes' quadrants. Although application contexts are suggestive of Pasteur- and Edison-type programmes, they may also require Bohr-type research for the reasons given earlier. In addition to those arguments, governments can choose to stimulate basic research out of cultural and intrinsic motives to sponsor one or more fields of research.

Stable agendas for applied research, together with financial and programmatic instruments, provide strong arguments for academic institutions with Pasteur-type research orientations to accommodate topics of those agendas in their research programmes. Especially longer-term issues, e.g. the *Societal Challenges* (European Commission, 2011), are a rich source for Pasteur-type, use-inspired basic research. It is interesting to note that, for such topics as e.g. climate change and sustainable energy, consistency between the different levels of government (regional, national, European, global) and between the different government agencies (science, environment, industry) is crucial to avoid fragmentation and inefficiency.

Other sources informing the applied research agenda are, of course, those linked to economic and industrial priorities, such as that in the Dutch *Key Industrial Sectors policy* (EL&I, 2011). Here we see that formerly distinct interactions, viz. research agenda-setting as part of the interaction between academic institutions and government, joint research programming as part of the interaction between (technical) academic institutions and industry, and industrial policymaking as part of the interaction between government and industry, become merged. The innovation process in this interaction typically has a cyclical character with many feedback and feedforward mechanisms that connect industry, government, and research, as in the model proposed by Berkhout et al. (2010). Such a tripartite ecosystem of interaction is often referred to as a *Triple Helix* (Etzkowitz & Leydesdorff, 1995), or more in particular in the Dutch context, a *Golden Triangle* (Lintzen

& Velzing, 2012). This development is closely linked to the industrial concept of *open innovation* (Chesbrough, 2003), in which industrial parties engage in joint research activities with each other and knowledge institutes, especially in the pre-competitive stages of research. Open innovation is especially useful in those contexts where the complexity of the challenges involved, and the need to overcome them, is considered greater than the benefits of exclusive access to the results of the research. This has understandably led to a wide spectrum of innovative legal constructions that regulate the exploitation of the intellectual property generated in such collaborations. Another phenomenon that is closely related to these developments is the notion of an entrepreneurial university. Although there is no authoritative definition of these concepts, they more or less stand for universities that have embraced new ways to foster the economic and social impact of their research and education, typically leading to new companies and enterprises as a result of university policies. In the Netherlands, the University of Twente became the first entrepreneurial university in the 1980s, mainly as a reaction to the then economic depression and the responsibility that was felt to do something about it (Boer & Drukker, 2011, pp. 146-149).

National governments have a tendency to reduce their role as financial sponsors of research in the presence of Triple Helix ecosystems, assuming that the presence of industrial parties will provide access to sufficient funding capabilities. Instead, they emphasize their role of regulatory architect and representative of the public interest. This is often based on an incomplete understanding of the risks and rewards involved in open innovations. The relevance of governments as sponsors of research for advanced technological developments is the topic of a book by Mariana Mazzucato, *The Entrepreneurial State* (2013). In one of her case studies involving Silicon Valley she shows that the successful innovations of Apple can be traced back to a plethora of federally sponsored projects by government institutions such as NSF, NIH, DARPA, DoD, etc. One of her conclusions is that such resources are essential at the stage where private companies cannot make large financial commitments because of the absence of a clear business case, and the government is needed in its entrepreneurial guise, as the underwriter of the risks involved. She also scolds the US government for having failed to design a framework in which the government not only reduces the exposure of private enterprises by funding the research, but would also share in the huge profits made out of the exploitation of the research outcomes. This is a relevant observation in a world where many companies have come to rely on public funding for long-term research.

Involving the public

One of the special features of the current Dutch National Research Agenda is that its conception was driven by a public consultation, i.e. the point of departure was a consultation of the general public on the questions that they would like to see addressed by Dutch research institutions. It is a characteristic of modern forms of governance that seek to involve the public more directly in the decision-making process. In this context, it is interesting to point out that the use of the internet has already impacted the public engagement with scientific research and innovation in very significant ways. As a platform the internet has given a huge boost to citizen science: with new ways to involve the public in data gathering and validation through *crowdsourcing* (e.g. Zooniverse, 2009); with ways of using networks of individual IT infrastructure as computational platforms (e.g. SETI); and with very impressive scientific contributions by amateur scientists, e.g. teenager Jack Andraka (BBC News, 2012). But the public has also become involved in matters of research and innovation policy through of *crowdfunding*, both of research projects (e.g. Experiment, 2014), and of ideas for innovative products and services (e.g. Kickstarter, 2009). Sometimes such instruments work together with traditional institutions, and sometimes they bypass them completely. It is to be expected that other new forms of such public engagement with research and science will develop in the time to come.

Given these developments, it seems clear that the format of the interaction between government, academia, and industry in innovation processes must be revised to incorporate these new forms of public engagement, leading to a *Quadruple Helix*, see e.g. Carayannis & Campbell (2009). That is, in reality we are moving from the arrangement in the form of a Triple Helix or Golden Triangle, to that of a square, as in Figure 5. In this new arrangement citizens contribute to the innovation process in their various roles as societal stakeholder, consumer, amateur scientist, investor, inventor, entrepreneur, etc.

This development clearly has consequences for all universities, and probably even more so for those with strong application-oriented profiles, for whom members of the public increasingly must be regarded as partners instead of mere clients for their outreach, education, research, or innovation programmes. It will be quite a challenge, both for the system as a whole and for the universities within it, to create the right level of responsiveness, being adaptive and flexible where this is needed, whilst still maintaining a clear agenda that warrants coherence, productivity, and continuity of the research and innovation processes.

Figure 5 Squaring the golden triangle

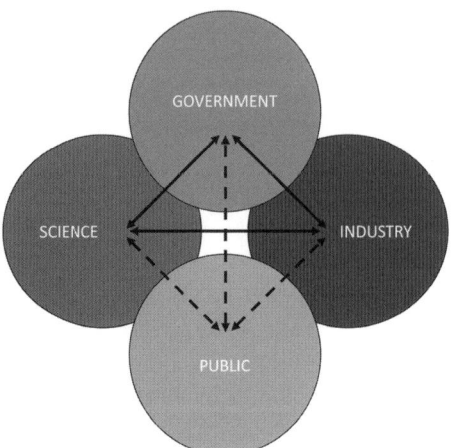

Conclusions

For universities of technology, and by extension other academic institutions whose research portfolio is substantially influenced by application domains, agendas of external organizations that are influential stakeholders in such domains, are natural points of reference for their research policies. Traditionally, national governments and industry have been such stakeholders for universities of technology.

I have revisited some of the ideas about the relationship between fundamental and applied research, especially promoting the views of Donald Stokes, who put forward that these are not exclusive categories, but rather that there is an important category of research that is both fundamental and applied, or *use-inspired basic research* in his terminology. The staple of research conducted at universities of technology belongs to this category. Moreover, Stokes pointed out that they form an important linking pin in the interaction between scientific knowledge and technological innovation. His account for the interaction between different types of research (Bohr: pure basic, Pasteur: use-inspired basic, and Edison: applied) provides a much better explanation for progress in science and innovation than the more traditional view of the linear model, largely due to Vannevar Bush, which unjustly positions basic research as the source of all innovation. Stokes' model, nevertheless, also implies a vital role for pure basic research (and pure applied research), by showing that productive innovation depends on the interplay between all types of research.

The models of Bush and Stokes pay little attention to the role of social sciences and humanities in the innovation process. It has become clear, however, that even in the context of technological innovation knowledge from these domains is most relevant for addressing challenges arising from the social and organisational embedding of technology-rich innovations, which could be characterized as *context-of-use-inspired research*. I have proposed a way to include its development as a new element in Stokes' dynamical model. Interestingly, here too the category of use-inspired basic research serves as the linking pin for interaction with other types of research.

Stokes' (extended) dynamical model implies that if one is interested in doing not only high-quality research, but also in contributing to the overall process of innovation, one should not only invest in the research itself, but also in its interaction with connected, other types of research. In considering the thematic links that define the interdisciplinary *routes* that have been identified in the Dutch National Research Agenda, it would also be good to see to what extent they also include the required types of research to be successful.

University profiles are an important interface between the activities of a university and its societal stakeholders, as characterization of its added value in the multiple networks to which it belongs (academic, research, education, regional, nation, global, industrial, societal, etc.). Although I have not spent many words on it, perhaps this is a good place to emphasize again that the profile should enforce consistency between the research and educational portfolios of a university, that is, there can be educational requirements for research that do not follow (directly) from intrinsic research and validation concerns.

Finally, I considered the interactions between academia and external stakeholders that influence the research and innovation agenda, most notably government and industry. This trilateral relation, or Golden Triangle, is evolving into an arrangement where citizens can also directly contribute to the interaction, often through *crowd-based* empowerment by internet platforms. This gives the public direct ways to influence research and innovation. It gives them, among other things, ways to select, participate in, contribute to, support, sponsor, etc. a growing range of research and innovation activities, adding a lot of dynamism and flexibility to the original arrangement. Since these developments are expected to grow considerably in volume and impact, they will produce *wisdom-of-the-crowd* generated additions and corrections to existing research and innovation agendas, both complicating and enriching the system.

In conclusion, one can say that a National Research Agenda is only one part in the complicated puzzle of interactions that determine the actual research and innovation processes taking place. For universities, and especially universities of technology with their explicit mission in innovation, successful research policy is an art of making the right connections: connections between Bohr, Pasteur, and Edison, between research and education, with the agendas of regional, national, and supranational government, with the priorities of industry, and, increasingly, with the preferences of the public. Real-world research policies, therefore, are determined by a multitude of concerns, in which the contents of the Dutch National Research Agenda, including its 16 preselected routes, can be helpful, but are not necessarily decisive.

Acknowledgement

The author would like to thank Erik van de Linde for a number of valuable suggestions.

References

BBC News, *US teen invents advanced cancer test using Google* (2012, August 20), retrieved from www.bbc.com

Berkhout, Guus, Dap Hartmann, and Paul Trott, 'Connecting technological capabilities with market needs using a cyclical model', *R&D Management,* 40 (5), 2010, pp. 474-490.

Boer, Jorrit de, and Jan Willem Drukker, *High Tech, Human Touch, A Concise History of the University of Twente* (Twente: University of Twente, 2011)

Brooks, Harvey, 'The Relationship between Science and Technology', *Research Policy,* 23, 1994, pp. 477-486.

Bush, Vannevar, *Science, The Endless Frontier, A Report to the President* (Washington, D.C.: US Government Printing Office, 1945)

Carayannis, Elias, and David F. Campbell, '"Mode 3" and "Quadruple Helix": Toward a 21st century fractal innovation ecosystem', *International Journal of Technology Management,* 46 (3), 2009, pp. 201-234

Chesbrough, Henry, *Open Innovation: The New Imperative for Creating and Profiting from Technology* (Boston: Harvard Business School Press, 2003)

EL&I (Ministry of Economic Affairs, Agriculture and Innovation), *Naar de Top: het bedrijvenbeleid in actie(s)* (The Hague: Rijksoverheid, 2011)

Etzkowitz, H., and L. Leydesdorff, 'The Triple Helix: University–Industry–Government Relations', *EASST Review, 14* (1), 1995, pp. 11-19.

European Commission, *Horizon 2020 – The Framework Programme for Research and Innovation*, 2011

Experiment, *Crowdfunding Platform for Experimental Research*, 2014, retrieved from experiment.com

Greenberg, Daniel, *Science, Money, and Politics: Political Triumph and Ethical Erosion* (Chicago and London: University of Chicago Press, 2001)

Hardy, Godfrey H., *A Mathematician's Apology* (Cambridge: Cambridge University Press, 1940)

Hodges, Andrew, *Alan Turing, The Enigma of Intelligence* (London: Unwin Paperbacks, 1983)

Kickstarter, *Crowdfunding site*, 2009, retrieved from kickstarter.com

Lintzen, Harry, and Evert-Jan Velzing, *Onderzoekscoordinatie in de gouden driehoek. Een geschiedenis* (The Hague: Rathenau Instituut, 2012)

Mazzucato, Mariana, *The Entrepreneurial State: debunking public vs. private sector myths* (London and New York: Anthem Press, 2013)

National Science Foundation, *Second Annual Report of the National Science Foundation FISCAL YEAR 1952* (Washington, D.C.: Goverment Printing Office, 1952)

Robbins, Herbert, 'Some aspects of the sequential design of experiments', *Bulletin of the American Mathematical Society, 58* (5), 1952, pp. 527-535.

Schumpeter, Joseph, *Theorie der Wirtschaftichen Entwicklung* (Berlin: Verlag von Dunkel und Hemblot, 1911)

SETI, *Search for Extra-Terrestrial Intelligence* (SETI@home), 2005, retrieved from setiathome.berkeley.edu

Simon, Herbert A., *The Sciences of the Artificial* (Boston: The MIT Press, 1969)

Stokes, Donald E., *Pasteur's Quadrant – Basic Science and Technological Innovation* (Washington, D.C.: Brookings institution Press, 1997)

Veerman, Cees, *Differentiëren in drievoud* (The Hague: Rijksoverheid, 2010)

Vincenti, Walter G., *What Engineers Know and How They Know It: Analytical Studies from Aeronautical History* (Baltimore: Johns Hopkins University Press, 1990)

Zachary, G. Pascal, *Endless Frontier: Vannevar Bush, Engineer of the American Century* (New York: The Free Press, 1997)

Zijlstra, Halbe, 'Speech at the STW Annual Congress 2011' (Nieuwegein, the Netherlands, 2011)

Zooniverse, *People-Powered Research* (2009), retrieved from: zooniverse.org

Too Big to Innovate?

The Sense and Nonsense of Big Programmatic Research

Brian Burgoon, Marieke de Goede, Marlies Glasius, and Eric Schliesser[1]

In our contribution to this volume, we argue that Dutch science-funding practices should be recalibrated because the status quo fails to meet its own stated objectives and is causing non-trivial harm along the way. We challenge, in particular, the existing bias toward identifying and awarding scholarly niches and national champions with large grants to ever tinier shares of the submitted proposals. We argue that this is wasteful spending and, when scrutinized, based on unrealistic assumptions about the nature of scientific research and the composition of the scientific community. The bias also skews the incentives for young researchers: by creating a culture of winners and losers, it demoralises promising young scholars and ends up mistakenly treating research and research impact as fundamentally opposed to teaching (rather than complementary activities). The result is that the existing system of funding may have the perverse, if unintended, effect of discouraging originality and innovation. The risk is that it undermines the 'culture of curiosity' that is essential to academic research.

We argue instead for a system of funding in which the existing pie is divided in a less bureaucratic fashion and among many more smaller grants, distributed among more researchers, so as to allow work in smaller, more fluid research combinations. We argue that this can also facilitate a more creative research culture in which different kinds of research approaches can be socially relevant, and in which research curiosity can flourish. Many of the arguments we offer here echo those made by others in various venues. But they are important to take seriously at this juncture in the development and scholarly soul-searching provoked by the National Research Agenda ('Nationale Wetenschapsagenda').

---

[1] All four authors are professors of Political Science at the University of Amsterdam, and have collectively raised grants from Dutch Scientific Council (NWO), European Research Council (ERC), Research Foundation Flanders (FWO), amongst others, worth more than 10 million euros. They have participated in many grant award committees.

Recent trends: centralized competition, declining success rates, increasing corporate orientation

The point of departure for our arguments are several observations about the trends and character of financing scholarship in the Netherlands and Europe, particularly of academic research in the social sciences and humanities. An important issue involves the actual level of financial support for research in the context of scarce research time, alongside teaching and administration. There is some debate as to whether, and in what realms, such financing has become less generous in recent years and decades if one looks at what various types of funding have been made available for actual research time and investment as opposed to various overheads (NRC/Rathenau articles and responses). It is, however, beyond dispute that research monies in Dutch academia have become substantially more subject to competition, as larger proportions of total funding have been shifted away from 'first flow of funds' investment (blocked and un-earmarked monies for research units and universities) and towards 'second ' (NWO) and 'third flow of funds' (EU and other sources) subject to individual or group, thematic or open, grant competitions.

While we would certainly join calls for more substantial investment in scholarly research, our concerns here are additional, involving three well-known features of this competitive financing. The first is that, particularly for the social sciences, acquiring research monies has become increasingly, and fiercely, competitive – in a way that leaves unfunded many researchers and projects deemed to be of high and fundable quality. This is certainly true with respect to the major funding sources for social science scholarship, the NWO, and the European Research Council. In the period 2009 to 2013, for which we have data, the average success rate for all science realms (NWO-Centraal, CW, STW, ALW, EW, Wotro, MaGW, GW, ZonMW, NORO, etc.) and all Dutch universities and institutes is 24%; for the humanities (GW) this figure is about 23%, but for the social sciences (MaGW) it is a mere 16% (NWO documentation 2015, via UvA Universitaire Onderzoekscommissie).

University administrators and Ministry of Education officials often point to the European Union as the promising funding source to take up the slack of national financing. Yet the competition for EU/ERC sources is even more intense, with financing and funding chances actually getting smaller – the average success rate for all funding lines (including sciences, medicine and humanities) has dropped from 20% to 14% in the recent *Horizon 2020* calls, compared to FP7 years. Also, financing in the EU's social-science realms has consistently seen the lowest funding rates, and hence been subject to the fiercest competition. In the previous FP7 structure, the 'Social Science and Humanities (SSH)', and 'Security' realms for all of Europe in the period up to 2013 amounted to 9% and 16.5%, respectively. In the new Horizon 2020 structure since 2013, 'Society' and 'Security' dropped even further to 8% and 11%, respectively, in 2014. Although Dutch universities have done somewhat better than average, our drop in success rates has in fact been greater, from 13% to 10% for 'Society', and from 23% to 12% for 'Security' in the same period. Although these figures reveal fierce competition the most recent trends are truly worrying, with the Horizon 2020 success rates for the social sciences dropping to a mere 4.2% (!) in 2015. While these figures are in and of themselves very worrying, they mask the fact that a great many (often older) researchers are only eligible for a small minority of grant lines, where the success rates are even worse (e.g. ERC Synergy grants had a success rate of 2.1% in 2014).

A second key characteristic in the implementation of research financing involves the focus on awarding winners and niches with a few large grants and 'consortia'. This is usually justified as recognizing the best research programmes, fostering national champions of excellence and taking advantage of economies of scale in research. The trends towards such champions involve not only individual multi-million-euro grants (e.g. VICI, ERC starter/consolidator/advanced) but also relatively new NWO instruments like the tens of millions of euros spent on single projects in the 'Gravitation' (*Zwaartekracht*) programme. To be sure, there is always a need to tie the financing of research to actual needs of projects, something that can require millions for ambitious research programmes – also in the social sciences and humanities. And there is a need to identify and encourage niches of research excellence within and between universities – something that NWO instruments may well be doing by inspiring productivity and some measures of quality among Veni, Vidi, Vici recipients (Gerritsen et al., 2013). Large grants might provide incentives to prepare and submit projects to compensate for the meagre chances of success. But this is a tendency that should be judged in light of the diminishing success rates and funding trends, meaning that there is a movement towards 'winner-take-all' dynamics where (growing) research demands and capacities are going unsupported.

Third, research financing includes increasing emphasis on more earmarked, thematically focused lines of research, where the themes are increasingly tied to manifesting or ensuring visible social and particularly economic relevance. This dynamic has long been true in the NWO and EU instruments. But it has become particularly clear in the transition from

FP7 to Horizon 2020 – where the latter puts greater weight on impact beyond the fundamental scholarly impact and where many research lines explicitly (and in practice) demand active collaboration with non-research-oriented entities in industry and civil society organisations.

In addition, and closer to home, the entire discussion of the Dutch National Research Agenda (NWA) and reform of the Netherlands Organisation for Scientific Research (NWO) has focused on reorganising funding lines into thematic areas, including the active promotion of 'top sectors' in the Dutch socio-economy. These trends have been reinforced by particular universities and research institutes within universities, such as the priority areas (with supplemental financing) identified by individual universities and faculties. This focus on themes is an important development to judge in and of itself, relative to the more open-ended focus of individual research grant lines; and it is important also to judge given the particular themes and kinds of partnerships that the NWA and top-sector policies envision.

Is the increase in competitive financing, clustering of research into major priority areas that pool facilities in large teams, and focusing on major themes of relevance leading to the innovative and internationally competitive scientific environment that many policymakers seem to dream of? Below we argue that the answer is 'no', for reasons that we divide into a discussion of the pitfalls of clustering into winner-take-all competitions and a discussion of the attempt to tie such clustering to particular themes of social and economic relevance.

The drawbacks of concentrating on big winners: small is beautiful

In this section, we first discuss some drawbacks of the current policy regime with its orientation toward awarding large research grants. We then offer an alternative vision in which we argue for a system that includes more and smaller research grants, selected and awarded through less cumbersome bureaucratic procedures.

In recent years, science policy and universities have championed changes to counter the model of individual, free, and unconstrained research in order to foster clustering and bundling. There is plenty to be said for this: perhaps the model of the lone genius, struggling to complete his (for the lone genius is nearly always gendered male) *magnum opus* in the proverbial attic room might no longer be the right model for young PhDs wishing to embark on a university career. As Stefan Collini (2012, p. 140) points out, 'scholarship is […] an inherently *cooperative* enterprise' (emphasis in original).

However, we argue that the consolidation of research funding into big grants, 'Gravitation' initiatives, and centres of excellence has reached its limits (Butterworth, 2015). It is both economically inefficient and demoralising to individual researchers. Explaining why this is so entails setting out three arguments. First, the costs of grant writing and reviews exceed the benefits with low success rates. Second, as grant size increases, it becomes less likely that research can afford to be genuinely risky and innovative. Third, the present system demoralises a new generation of excellent researchers before their careers even get off the ground.

First, the costs of grant writing and reviewing are reaching a limit where they are disproportionate to the payoff.[2] In our own experience, this has become particularly acute as a problem given the *opportunity costs* of grant work, the weeks and months (and accompanying stress) that could otherwise be spent on one's actual scholarship and output. And such opportunity costs are higher particularly where one must tailor new proposed lines of research and collaborative organisation to suit the vagaries of particular calls for proposal in a thematic grant line. These are more obviously wasted efforts should a proposal not be granted. Additionally, a host of referees and grant committee members are spending *their* time reviewing mostly unsuccessful grant proposals instead of doing their own research.

These are perennial worries about competitive review in a winner-take-all setting, but evidence from outside of Europe make this problem even greater. In their study of the costs of the grant peer review system in Canada, Gordon and Poulin (2009) found that the cost of preparing and reviewing grant applications now exceeds the gains of selection. They argue that it would be cheaper and more effective to distribute small, direct grants without peer review to all qualified researchers. They consider the grant competition system to be skewed, not just because grant-giving bodies often have near 'monopoly status' (ibid., p. 21), but also because they need to compare 'competing worthiness of distinct goals', rather than adjudicating between 'people trying to attain the same specific goal' (ibid., p. 16).

In addition, there is solid empirical evidence of diminishing returns of grant size. A recent study shows that '[r]esearchers who received additional funds from a second federal granting council, the Canadian Institutes for Health Research, were not more productive than those who received only

---

2   For a useful, styled arithmetic exercise of the waste in the current Dutch grants system, see De Cruz (2014); for published research on opportunity costs in grant writing with data from the US, see Von Hippel & Von Hippel (2015). Australian researchers have also found non-trivial impact on emotional wellbeing of researchers, see Herbert et al. (2014).

National Science and Engineering Research Council (NSERC) funding. Impact was generally a decelerating function of funding. Impact per dollar was therefore lower for large grant-holders' (Fortan and Currie, 2013). There is an intuitive reason behind this result: as funding increases, excellent researchers are turned into bureaucrats who must manage other people and spend increasing time on reporting rather than on research.

Gordon and Poulin argue that it might be better to distribute grant money randomly or to spread it equally. The outcomes of grant evaluation procedures are often compared to the outcomes of a lottery, and judged to be 'random and arbitrary' (2009, p. 21). But this metaphor does not do justice to the hard work and serious effort by all participants in the grant review procedure. Research councils take great care in designing procedures that are clear and fair, given serious constraints on their budgets.[3] Based on our own experiences with participating in grant awarding and grant review work, we posit that the outcomes of grant review are less like a lottery and more like a carefully polished funnel. Often, grant competitions – despite all their goals of excellence – support compromise and middle ground. A layered collation of assessments underpins any grant decision: for example, in a typical NWO Vidi competition, at least two pre-reviewers (members of the committee) will assess the proposal; then (if the applicant is lucky not to be rejected at pre-review stage) at least four external reviewers (sometimes six or seven) assess the proposal; then the whole committee of twelve to fifteen or more members assess and rank the applicants' interview performance; finally, the NWO domain chair has to formally approve the nominations. At all these stages – except perhaps the last – it is important that the proposal receives support and instils enthusiasm with reviewers and committee members. But it is equally important that, at all these stages, the proposal does not challenge or alienate its readers, or provoke strong negative reactions. All other things being equal, unconventional and controversial approaches within a discipline fare less well than standard and safe approaches. The multilayered and reiterative review system adds up to support mainstream and incremental proposals, not necessarily originality and excellence.

Second, then, we argue that the large grant competitions are inherently conservative in the outcome patterns they generate. This is partly due to the way innovation in research, perhaps particularly in social sciences and humanities, emerges not so much or only from economies of scale but from

---

3   Whether grant review procedures are clear and fair is beyond the scope of this chapter. Much work is done, for example, in NWO's so-called 'pre-advice' forms, which remain entirely obscure to applicants.

'economies of scope' (see Teese, 1980). These can be understood as informal, intellectual 'trading zones' where theoretical and empirical insights developed with respect to one research line spilling over to others (see Galison, 1997, writing about physics). Institutions or research communities encompassing scholars from very different theoretical and methodological traditions – and, indeed, different disciplines – can trade insights in competitive or collaborative dialogue, even while each is working alone on his or her own boutique research programme. This can inspire innovation and creativity much more than does the pursuit of scale economies. Harnessing such gains from diversity argues against the identification of large-scale clusters or niches.

In any event, the privileging of big priority winners can be conservative and may fail to support innovation given the evaluation procedures governing the picking of winners (and losers). An important example is that feasibility is often an explicit, non-trivial evaluative criterion. To write a grant proposal able to survive the rigorous review procedures that are – necessarily – designed for the largest of grants, the applicant needs to be fully immersed in the subject matter: s/he needs to be thoroughly familiar with the literatures and debates, know exactly what s/he wants to examine, why and how. Successful proposals have to articulate what PhD candidates will be doing in three to four years' time, where they will go, who (for example) they will interview, and what they will ask. The expected outcome of the proposed project and what major breakthroughs it is likely to deliver need to be specified in advance. It is entirely understandable that grant-giving bodies, handing out millions of public money, should desire this level of detail. But if all uncertainty and possibility of surprise is eliminated, why would this process lead to innovative and creative projects?

Moreover, the emphasis on big grants has an anti-innovative effect on the development of young scholars doing their PhDs. Whereas the doctorate was originally considered to be a young scholar's 'master proof', demonstrating his or her ability to conceive, carry out, and write his or her own research from start to finish, we are now training a generation of scholars whose first extensive research experience is in carrying out a research project formulated by someone else. The innovative potential of research questions formulated by graduates in their twenties is largely getting lost.

Large individual grants are the main funding instruments through which creative, independent and curiosity-driven research is supported. However, their evaluation procedures create incentives for applicants to continue with research that has a status quo bias built into it. In addition, existing track records within a given research area will be a major evaluation criterion.

This raises barriers to innovation for even the most successful researcher. Thus, while the NWO experience might entail large grant competitions promoting productivity (Gerritsen et al., 2014), large grant competitions can be expected to 'lock scientists into narrow paths [...] reducing the adventure, innovation and scope of their discovery' (Gordon and Poulin, 2009, p. 20).

Within a grant review system in which outliers are systematically disadvantaged, what happens to surprise, to curiosity, to adventure? As Patricia Pisters (2015) has asked, are big grant competitions sufficiently able to support 'unexpected connections [and] unpredictable discoveries'? Do they succeed in stimulating 'the human avidity to know', described by Foucault (1989, p. 305) as an ethos of curiosity that has 'a readiness to find strange and singular what surrounds us; a certain relentlessness to break up our familiarities and to regard otherwise the same things'? Not all good research might know its outcomes in advance.

Third, and perhaps most importantly, we now risk creating a generation of very good and very disappointed young scholars. In the social sciences, direct funding for PhDs has largely dried up. That means that aspiring PhD students are faced with a choice: either they apply to do a PhD within a senior scholar's funded project, which means they can pursue their individual intellectual curiosity only in limited ways, or they pin their hopes on the NWO Talent scheme with a less than 10% success rate and a process that takes nine months, or they commit to doing an unfunded PhD, whilst making a living with teaching or other work.

Once in possession of a PhD, academics are once again confronted with the two-tier system. In many universities, at least in the social sciences and humanities, a full teaching load leaves too little time to maintain an internationally visible research career, however much one loves teaching (and many successful researchers are also passionate teachers). Most Dutch universities lack a system of regular research sabbaticals. This means that young lecturers strongly feel a need to bring in grant income, not just to achieve tenure, but also to shield their research time from the pressures of teaching and management. Now that success rates in the most important grant competitions have fallen so low, young researchers have to get accustomed to being rejected before their careers even get off the ground properly. Clearly, dealing with rejection is part of academic life, and in some cases it leads to better proposals and more determined researchers. However, our research funding system stimulates profound competitiveness with very small chances of winning the competition. Promising young researchers face increasingly pressurised environments, because while grant success rates are going down, the sense of the importance of grant success to their career prospects is going up. And the problem is compounded for older researchers who survive in such environments – as for them there are very few funding lines to even compete for, making the fundraising standard of success and quality all the more difficult to meet. There is little justification for this state of affairs; people with roughly equal educations and productivity levels are treated as if they have extremely different research skills.

Our proposal

There is no doubt that the future success of grant competitions and the legitimacy of the research councils requires a significant increase in success rates. We do not deny that there *is* a future for grant competitions and research councils. Letting universities distribute all the monies (as Gordon and Poulin suggest) is not a solution: it would increase internal competition, and possibly lead to obscure decision-making (by university managers rather than academic peers). Basically, success rates can be improved in two complementary ways. The first is a very substantial increase in and diversification of government funding for research across the board, including PhD projects, small grants, and funding for large collaborative project. We certainly support such an increase. But the second way is crucial to the current climate where more generous funding appears politically unlikely: success rates can be increased by developing more varied competitions for many, smaller grants and smaller consortia with less burdensome application and review criteria, including periodic small-scale grants for researchers in good academic standing to support, say, modest periods of leave or research assistance. This simultaneously broadens eligibility criteria, because it would open competitions to many more ages and categories of researchers in academia. And it could include more funding for individual PhD projects, allowing future PhD candidates to write their own original proposals (currently, the NWO PhD grant competition *Onderzoekstalent* is one of the worst when it comes to success rates).

Research funding should do more to stimulate independent research and smaller-scale projects (Pisters, 2015). In many research lines in social sciences and humanities, valuable research can be carried out with grants that run into the thousands and ten thousands, rather than millions, of euros, funding some months of teaching buy-out and some travel, research assistance, or data purchase.

Finally, the current funding system only recognizes individual 'principal investigators' who are expected to hire PhDs and post-doctoral researchers, and large consortia with multiple teams working together. But social scientists and humanists typically collaborate in very small, often horizontal teams of 2-4 people, sometimes based at the same university but often not. Funding these kinds of collaboration, again with small funds and low-intensity procedures, would better connect funding opportunities to actual research practices, instead of getting the practices to contort themselves to be in conformity with eligibility criteria.

Better success rates at research grant competitions entail a better balance between the investments in grant writing and reviewing and the payoffs; it means more room for adventurous, curiosity-driven research (in addition to large projects); and it provides more stimulus and chances to a wider group of young researches.

'Knowledge utilization' in the service of business and government

In this section, we chart how Dutch science policies have come to translate the need for science to be socially relevant into a demand that it should directly serve the corporate sector or the government's knowledge needs. We then outline our own vision of scientific research as networked into, feeding on, and serving a knowledge society, and the kind of funding strategy that would befit and benefit this vision.

More than sixty years ago, the Dutch government founded the Netherlands Organisation for Pure Scientific Research (ZWO). Its remit was to exclusively fund non-applied research. In 1988, the organisation dropped the term 'pure' and began to fund 'both curiosity-driven research and research into issues that occupy [sic] the world.'[4] Social relevance has been an – initially optional – criterion for assessing its research proposals ever since, one that has animated social scientists, since social trends and problems are their object of research, making it inherently relevant to society.

Recently, this element of assessment has been relabelled 'knowledge utilization', reflecting the insight that it is not enough for research just to be relevant to society in principle, but that efforts need to be made for social actors to be able to understand and utilize research findings. In itself, this shift is to be commended: scientists should not be content to publish only in specialist journals and leave a special class of knowledge entrepreneurs to take up their findings – or not. Funded by the taxpayer's money, they should make an effort to explain what they do and why it matters to social actors

---

4 See NWO's mission statement: www.nwo.nl/en/about-nwo/mission+and+vision

who may learn and profit from their findings. But in NWO competitions, the ways in which knowledge utilization is identified and assessed is sometimes unclear: should researchers blog, tweet, and write op-eds, should they offer direct policy advice, should they contribute to economic growth? Or all of the above? In addition, there is a risk that valorisation prioritizes economic utility and downplays cooperation with the social sector, including NGOs and civic groups.

This upgrading of the old 'social relevance' criterion is part of a broader international trend. In the United Kingdom, the latest national research assessment, Research Excellence Framework, now includes a criterion on impact, which requires institutions to submit case studies documenting how research has had an 'impact', defined as 'change or benefit to the economy, society, culture, public policy or services, health, the environment or quality of life, beyond academia' (REF, 2014, p. 6). The emphasis on impact is problematic on various levels, but at least the referent of the impact is very broadly defined.

In the Netherlands, by contrast, we have recently witnessed a much narrower interpretation of what constitutes appropriate social impact. Ten years ago, policymakers introduced the idea of 'valorisation' as an aim that universities ought to pursue, meaning 'turning research results into economic value'. More recently, in 2011 former Economic Affairs Minister Maxime Verhagen launched the catchphrase 'kennis – kunde –kassa' (knowledge = skills = cash) to express his vision of the contribution of science to society. Investments in science, in other words, were to be translated directly into an increase in the profit margins of the corporate sector. In practical terms, this vision was translated into generous support for the above-mentioned 'top sectors': collaboration between academia and Dutch corporations in nine specific sectors.[5]

While Verhagen's vision may have been extreme in the candidness with which it reduced the purpose of scientific endeavour to the fattening of corporate calves, it is again part of a broader European trend in seeing science as an engine for innovations with economic benefit. The European Research Council, one of the EU's primary funding instruments, appears at first sight very different, with an emphasis on 'investigator-driven frontier research' and a recognition 'that research at and beyond the frontiers of understanding is an intrinsically risky venture'. Yet it also insists that such

---

5   See Wetenschappelijke Raad voor de Regering [Netherlands Scientific Council for Government Policy], *Naar een lerende economie*, Report No. 90, November 2013, for a critical assessment of the top sector policy even from the perspective of its stated aim of serving the Dutch economy.

research must be 'of critical importance to economic and social welfare'. The economic element is clearly privileged, as illustrated by the ERC Proof of Concept grant available to existing ERC grant recipients for 'bridging the gap between research and a marketable innovation'.[6] There is no equivalent grant for translating one's research findings into social benefits.

The current Minister of Education has walked away from the knowledge-skills-cash catchphrase (characterizing it as 'revolting'), but the tendency to equate social actors with corporate actors remains unchanged. The knowledge coalition behind the Dutch National Research Agenda that is the subject of this volume consists of a wide variety of research institutions, and just two social actors: the Confederation of Netherlands Industry and Employers (VNO-NCW) and the Netherlands organisation for small and medium enterprises (MKB). It is as if the knowledge needs of society are wholly reduced to being factors of economic production.

Funding opportunities for other types of partnership tend to be very narrow and directed. To give one example, a current call for applications by NWO on Security & Rule of Law in Fragile and Conflict-Affected Settings initially appeared relevant to the research of some colleagues. However, it turned out that the research could only relate to specifically named countries where the Netherlands is active as a donor. Hence, one colleague who works closely with Médecins Sans Frontières (MSF) on the influx of refugees from the Mediterranean could not apply because the refugees have fled the named countries, while another colleague who works on precisely the right issues in Latin America could not apply because the region, no longer funded by Dutch development aid, fell outside the call's remit.

Knowledge utilization in the service of a knowledge society

It is appropriate that government funders should encourage scholars to make their work directly relevant and available to social actors. But the variety of ways in which social scientists are already engaging with 'societal stakeholders' is greater than funding agencies can possibly imagine. In our direct environment, we witness extreme variety in the type of actors scholars engage with and the depth, length, and scale of engagement. In terms of the type of actors, some of us advise central bankers and European policymakers, whereas others advise disadvantaged schools, people living with HIV Aids or environmental activists. The depth and length of our

6   See ERC website: https://erc.europa.eu/proof-concept

engagement varies from research projects co-designed from beginning to end with a social actor, such as the Knowledge Programmes initiated by development organisation HIVOS, or a public mediation programme where research and practice are intertwined, to five-minute radio interviews interpreting the latest election polls. The scale of our engagement varies from very targeted interventions such as expert testimony before specific national or European committees, to media performances, columns, or blogs intended for the general public.

We propose that, in addition to the existing practice of asking grant-seekers to describe their plans for knowledge utilization, a set percentage of national and supranational funding should be set aside for research involving in-depth collaboration between researchers and social partners. The forms this may take and the type of societal stakeholder that could be involved should remain largely open. NWO and other funding bodies could build on the NWA exercise in fostering engagement between academics and society at large by creating a pool of volunteer lay reviewers from all sections of society to review such collaborative proposals alongside academic peers. As with the other forms of funding we propose, grants should be small and multipurpose, and procedures should be light. An emphasis on small grants, for example up to €50,000, will not only have the advantages sketched above, but also prevent capture of the scheme by big corporate interests.[7]

Such a scheme would exemplify a funding policy that prioritizes knowledge utilization *without* steering it towards particular (corporate) actors, particular (government policy) agendas, or particular notions of productivity, whilst neglecting or stifling many others. It is in line with what Schnabel et al. have characterized as the 'network university' that serves not just a knowledge economy, but a knowledge society (Sociale Wetenschappen, 2014, pp. 55-56).

Finally, funding bodies should explicitly recognize the most obvious and natural way in which scholars translate their research work into broader social knowledge: via the classroom. Year in, year out, social scientists teach new generations of future societal leaders and citizens what they have learned through their own research and that of others. Once science policymakers recognize this, we can stop treating research and education as opposed to each other.

---

7   From an economic perspective, this situation resembles a form of rent seeking by richly endowed and well-connected corporate agents, who should, in fact, be able to fund profitable research without government aid. It is by no means obvious that the existing funding policies are the best way to increase social goods.

Conclusion

We have argued against the existing bias toward awarding large research grants, which, given the size of the current research pie, generates extremely low success rates, cumbersome bureaucratic procedures, and considerable opportunity costs. In addition, the bias toward large research grants encourages less innovative research and thereby fails to produce the intended policy goal. We believe that the available grant mix should be diversified with increased availability of smaller grants that can be awarded to more members of the research community. In addition, we have argued that research impact and utilization should be oriented not just toward well-connected corporate agents, but toward a wide diversity of societal stakeholders, including those found in classrooms.

References

Bollen and Scheffer https://omslag.de/onderzoeksfinanciering/de-wijze-massa/

De Cruz, H., 'Gedijt wetenschap het beste bij competitie? Reflecties op de wetenschapsvisie', 12 december, 2014, *Bijnaderinzien,* http://bijnaderinzien.org/2014/12/14/gedijt-wetenschap-het-beste-bij-competitie-reflecties-op-de-wetenschapsvisie/

Fortin, J.-M., and D.J. Currie, 'Big Science vs. Little Science: How Scientific Impact Scales with Funding', *PLoS ONE* 8(6), 2013, e65263. doi:10.1371/journal.pone.0065263

Foucault, Michel, 'The Masked Philosopher' in *Foucault Live: Collected Interviews 1961-1984*, edited by Sylvère Lotringer (New York: Semiotext(e), 1989 [1980])

Galison, P., 'Image & logic: A material culture of microphysics' (Chicago: The University of Chicago Press, 1997)

Gerritsen, S., E. Plug, and K. van der Waal, 'Up or Out?: How individual research grants affect academic careers in the Netherlands', *CPB Working Paper No. 249*, 2013, pp. 1-33

Gordon, R., and B. Poulin, 'Cost of the NSERC Science Grant Peer Review System Exceeds the Cost of Giving Every Qualified Researcher a Baseline Grant', *Accountability in Research: Policy and Quality Assurance* 16 (1), 2009, pp. 13-40

Herbert, Danielle L, et al., 'Public health – Research: The impact of funding deadlines on personal workloads, stress and family relationships: a qualitative study of Australian researchers' *BMJ Open*, 4:3, 2014, e004462. doi:10.1136/bmjopen-2013-004462

Pisters, Patricia (2015) http://rethinkuva.org/blog/2015/03/16/jams-loops-and-downward-spirals-in-the-academic-system/

Research Excellence Framework (REF): *The results*, Higher Education Funding Council for England (HEFCE), 2014

Sociale Wetenschappen: 'Verantwoord en Verantwoordelijk, Sectorplan Sociale Wetenschappen 2014', Commissie Sectorplan Sociale Wetenschappen, 2014

Teece, D.J., 'Economies of scope and the scope of the enterprise', *Journal of economic behavior & organization*, 1(3), 1980, pp. 223-247

von Hippel T., and C. von Hippel, 'To Apply or Not to Apply: A Survey Analysis of Grant Writing Costs and Benefits', *PLoS ONE* 10(3), 2015, e0118494. doi:10.1371/journal.pone.0118494

Werfhorst, van der, Herman, 'NWO benadeelt de sociale wetenschappen', 15 March, 2015 www.hermanvandewerfhorst.socsci.uva.nl/blog/science/nwo/

The Art of Asking Questions, and why Scientists Are Better at it

Herman van de Werfhorst[1]

Introduction

As explained in other parts of this volume, the Dutch government has involved the public in generating 'questions for research'. Through the National Research Agenda (*Nationale Wetenschapsagenda*, or NWA) individuals and organisations were invited to pose questions for scientific research. A large number of questions have been formulated, across all disciplines, and of varying forms. By inviting the whole of society to ask scientific questions, the principal aim is to 'solve problems'. But another important purpose is to enlarge the legitimacy of academic research. If society can influence the agenda of academe, it will be easier to defend that researchers do basic research without a direct 'return' in the form of economic or social spin-off. We can let scientists play, but according to the rules set by society. Even if the legitimacy issue is not at the core of the matter, the structuring capacity of the NWA for the research agendas of tomorrow does pose the science–society connection at the heart of the endeavour. We thus need to see whether the problem-solving and legitimizing ambitions of the NWA are achieved in the current process.

While I share the view that it is important that scientific research finds legitimation in society, and I am all for solving the problems that emerge in society, I fear that the way society and research have become interconnected in the current process is ineffective. More specifically, I have three worrying questions about whether and how the more relevant scientific research can be produced in the way the NWA is set up. First and foremost, is it sensible to let society do the job, by letting it ask questions? Or would there have been another, more effective way to improve the connections between research and society? Letting society do the asking, letting the public, firms, and interest groups take the initiative in the agenda-setting, is, in my view, worrisome. It invalidates one core quality of scientists, that they master the art of asking questions better than anyone else.

---

1 Herman van de Werfhorst is Professor of Sociology at the University of Amsterdam and director of the Amsterdam Centre for Inequality Studies (AMCIS).

Second, will legitimacy of scientific research be enlarged if the public can formulate questions? What can we learn from other areas where the public directly influences agenda-setting, in particular politics? And third, is it the public where the connection with society should be sought, or had we better seek it elsewhere? Can the public oversee the various solutions that scientific research can achieve?

The direction of influence: who asks the best questions?

The approach taken in the NWA is that society can formulate questions for science. This has resulted in almost 12,000 questions in various shapes and forms, varying from points on the horizon ('how can we make society more XX in the 21st century?') to proper research questions. These 12,000 questions have been summarised in 140 'cluster questions', again in various shapes and forms. Scientists and knowledge institutes have played an important role in getting from the 12,000 questions to the 140 cluster questions.

My first worry is that the direction of influence, where society influences the questions that scientists ask, is the wrong one. I agree that it would be good to stimulate interaction between scientists and society. Possibly scientists have had too little focus on the usefulness of their expertise for business, technological, and social issues, and the aim to bridge scientific expertise with partners in the field, such as businesses, governments, or other stakeholders, is laudable. Yet letting society do the asking is a mistake.

Formulating research problems is at the core of the scientific process. Research questions guide our work. To formulate them properly is a skill in itself, a skill that takes more than requesting solutions for everyday problems. A good research problem is not just a guide for looking for facts. A good research problem is informed by, and grounded in, scientific theories. Answering them helps to better understand the merits of these theories and, thus, to improve our knowledge of the world. Moreover, as examples from my field (sociology) illustrate, research problems are improved if they are layered: a specific research question can be seen as a sub-question under a broader research problem. The whole field of sociology can be subsumed under three overarching problems, according to Ultee, Arts, and Flap (1996): inequality, social cohesion, and rationalization, or, according to Wilterdink and Van Heerikhuizen (2013), under four types of social relationships (economic ties, political ties, affective ties, and cognitive ties). Independent of which approach one prefers, it is crucial for scientific progress, also to the aim of solving problems, to formulate problems that are believed to be of broader scientific interest for a discipline.

Part of the challenge of formulating research problems is the delicate balance between problems and theories – a balance that will not be at the forefront of societal stakeholders posing questions to us. Somewhat jokingly we sometimes hear that scientists come in two sorts: those with a problem looking for a theory, and those with a theory looking for a problem. This distinction is, however, hardly useful, and ill-informed by a Kuhnian perspective emphasizing that scientific problems are asked within the context of theoretical paradigms. Problems that fall out of the blue may be looking for a theory, but if they are not posed from the interest of a particular theory we are left with fact-finding rather than theory development. Hypotheses can be loosely formulated but if their test doesn't say anything about a broader theory we have gained little relevant knowledge. Hypotheses should, therefore, not 'come from the neighbour' but rather be developed from the perspective of (layered) theories. On the other hand, if a theory is looking for a problem we may end up with research that is hardly connected to the real-world problems for which scientific knowledge is useful. If, for instance, we are interested in broad theories that say that humans are altruistic by nature, we may end up with some interesting and well-done laboratory experiments, but without clear linkage to the real-world problems in which altruism and cooperation may be decisive. In short, only by close interaction between problems and theories can scientific research emerge that is able to help solve real-world problems. But it is doubtful whether the one-way street of asking questions as employed in the NWA is able to improve this interaction (notwithstanding that scientists have been involved in the classification of questions).

The good thing about the NWA is that it promotes a closer interaction between science and society to solve real-world problems. Such an interaction will not happen automatically; scientists cherish their academic freedom, and theoretically constrained problem formulations lead to scientific progress. But are scientists really so distanced from real-world problems, and from the applicability of their theories? I don't have that impression, and criticisms we sometimes hear from politicians that we should leave the ivory tower are misplaced. The problem is not that scientists refuse to descend from the tower. Rather, the problem is that partners in society are not willing to posit their specific problems within the context of broader scientific theories. It is not a lack of noise; it is a lack of audibility.

As an example, I would like to take the reader to my field of education research, cross-cutting between the fields of sociology, education, and economics. The strong empirical focus of researchers in this field ensures that many of us almost automatically think in terms of applications. While some discussion emerges about the external validity of experimental research, and about causality in non-experimental research, there is no unwillingness to be involved with the field (schools, policymakers). What seems to be a bigger problem is that non-academic partners have a strong influence on the funding of educational research in the Netherlands, so that sometimes scientifically excellent proposals do not get funded, even if the ensuing knowledge would find applications in the field.

Nevertheless, in the field of education and elsewhere, scientists can improve making their case for the applicability of their knowledge to real-life problems, and an improvement of the interaction between non-academic partners and scientists is desired. However, if science is to be more strongly connected to the solution of real-world problems, and if we believe that research problems are only scientifically valuable if they are developed in close connection to theories, I would think that scientists should have the first hand in the game. A more effective way to promote interaction would be to stimulate scientific researchers to help partner organisations to *formulate* research questions. By involving scientists in the formulation of research problems – not only their own research problems but particularly the research problems of 'society' – practical problems can be placed within broader theoretical agendas that can be overseen by scientists. This implies that mundane real-life problems become scientifically relevant, which further ensures that the problems will help to improve our understanding of the world. Through better theories we can solve problems, not because one particular acute problem emerges but because each particular acute problem is part of a larger scientific challenge. And scientists are better able to see that.

Looking at the cluster questions

Looking at the 140 cluster questions, we see that the nature of the questions differs a lot, varying from purely scientific research problems, to a mixture of research and societal challenge, to clear societal challenges without a clear research agenda emerging. Societal challenges are typically practical questions about the future: 'How can we ensure that... ?' Scientific problems aim to find explanations for existing (or past) phenomena: 'How can we explain...?' For instance, cluster question 11 (*How can we manage water carefully in the future?*) is, by nature, a societal challenge more than anything else. There is no clear scientific research problem stated in the question. Another example of a policy-driven (rather than scientific) question is number 94: *How can we improve healthcare, but at the same time keep it affordable?* These are in fact two societal challenges, rather than a scientific problem. Other cluster questions combine scientific questions with societal challenges, such as 43: *What are the causes and consequences of migration and how can we deal with it?* Although the latter part of the question is not quite clear to me, it illustrates a societal challenge rather than a scientific problem, while the study of causes and consequences of migration clearly relates to a proper and relevant research agenda. Another example where scientific and societal challenges are combined is question 108: *Which social changes caused by technological changes can be expected, and affect our wealth?* The confusing part is 'can be expected'; one thing we have learned is that social sciences can poorly predict the future, but is better at explaining the empirical observations of the present or the past. But besides that, a clear research agenda emerges about the interrelationship between social and technological changes and wealth. Another clear scientific problem emerges in cluster question 31: *What does globalization mean for our cultural identity and the determination of the position of the Netherlands in the world?* Especially the first part of the question can easily culminate in a relevant research agenda. Thus, some questions are more easily seen as building blocks of a research agenda than others.

An important exercise of the NWA is furthermore to provide a limited number of 'exemplary routes' through the 140 cluster questions. As the term illustrates, these routes are examples, and could be extended by other routes that scientists or stakeholders can create through the cluster questions. In fact, establishing routes can be seen as an important way in which scientists can categorize cluster questions into layers of a larger scientific problem; a main criterion for scientific relevance, as I have laid out above. However, the Dutch ministers of Education, Culture and Science, and of Economic Affairs, have written to Parliament that the current routes will be used as an anchor for science policy, by using them as building blocks for research funding of the Netherlands Organisation for Scientific Research. Before we know it the routes have become a reality, while it takes a more thorough involvement of academics to see all the relevant layers in the 140 cluster questions. From my expertise it is, for instance, remarkable that there is no route for youth and education (including cluster questions from psychology, social sciences, and health), or for life courses (combining clusters from economics, health, and social sciences), or for diversities and inequalities (social sciences, health, economics, psychology, political science, philosophy). I am sure scientists from other fields will find similar omissions in the current list of sixteen routes. The point is that the 140 cluster questions have a broad coverage across scientific fields, and many fields will feel rather well-represented (although not always by clear research problems). Nevertheless, the routes seem rather arbitrary, and they should not become building blocks for policy without a stronger involvement of scientists. If scientists are better at asking, and layering, questions, it should be scientists who determine the routes through the cluster questions.

Will involvement promote the legitimation of science?

The connection between science and society is not only relevant from the perspective of solving problems, the main focus of the NWA, but also from the perspective of legitimacy of science. My second worry is that a stronger involvement of societal stakeholders in the scientific agenda-setting will not automatically improve the legitimation of science.

Empirical research shows that science, and especially scientific institutions, are not always trusted by the public (Achterberg, 2015; Achterberg et al., 2015). Especially the lower-educated population distrusts scientific institutions, as opposed to the more highly educated population. Importantly, the educational gradient in trust in science is explained by cultural discontent with the complexities of the modern social order, where more uncertainty and 'anomie' (normlessness) are experienced by the less educated. It should be noted that the lower educated are more distrustful of all institutions. Moreover, overall the trust in science is highest of all known institutions, including the legal system, medical doctors, and politics.

It would require more empirical research than currently possible to be sure, but it is very likely that the public that have been involved in generating questions for science covers the well-educated fraction of Dutch society (and the organisations that have posed questions are also populated with more highly educated individuals). So, legitimacy is increased among the group that already puts strong trust in science, which may in fact *increase* the social differentiation in trust in science. With regard to trust in institutions (be it scientific or other institutions such as Parliament, the police, or the legal system), one may claim that social cohesion in society is particularly enlarged if there is little variation in trust across social and demographic groups (Green et al., 2006). So whether the NWA has improved social cohesion by enlarging the legitimacy of science can be questioned.

It is interesting to see how public involvement in an institution is related to trust in this institution by looking at the field of politics. In democracies, the public elects Parliament – a closer connection between public involvement and national institutions is hardly possible. Yet, in the Netherlands only half of the population trusts Parliament, a portion that is, moreover, decreasing (Dekker and De Ridder, 2015). So there is no clear relationship between potential involvement and trust. It is therefore unlikely that a stronger involvement of the public in scientific agenda-setting will improve the legitimacy of scientific research.

Who can judge whether society has improved?

The third and final worry concerning the procedure of asking society to pose questions concerns a more fundamental issue relating to the enhancement of the legitimizing and problem-solving capacities of the NWA. The question is, who is able to see the benefits of scientific research? A quest for solutions of problems does not only require that we generate research problems, but also that we know which problems are already solved at various levels. And while scientists and other partner organisations together may get rather far in deciding which solutions are still desired, it is doubtful whether laymen can be of much help here.

Thinking about the 'problems' that science can solve, these problems come in various forms. Technological innovations may help businesses, governments, schools, and civic organisations. Clear problems may emerge in terms of, say, sales, water management, ICT in schools, or increasing membership of non-governmental organisations, and technology may help to solve them. What Mazzucato (2013) shows is that technology is often funded by the state through fundamental research, without partner involvement from the business community. The most prominent example is the iPhone, many parts of which have been developed with funding from the National Science Foundation in the United States.

But problems are not always technological. What about knowledge of the history of monotheistic religions, or of international relations; do we believe that the public can oversee the problems that need solutions? Or more directly related to my field, could inequality of educational opportunity in Western societies exist without people being aware of it, and/or without people being worried about it? It is striking to realise that the Dutch government thinks that everybody has equal opportunities in Dutch education; it is believed by many policymakers that if people have the abilities and the motivation they can achieve all levels of education. It takes persistence of researchers, without liaison with policymakers, to show that parental background still matters for the (binding) advice that school teachers give to pupils concerning their secondary school type, even when controlling for intelligence and standardized test results (Van de Werfhorst et al., 2015). If we had to rely on policymakers or politics, it is unlikely that this knowledge would have been presented. Likewise, early-selecting systems of education have been shown to be related to larger inequalities of opportunity, especially in the absence of standardized tests (Bol et al., 2014). Given the fact that the education field is currently allergic to 'educational system questions', it is unlikely that the theme of early selection would have been put on the agenda if we had to rely on partners in the field in the formulation of questions.

In short, both with regard to problem-solving and legitimation, it is doubtful whether societal stakeholders or the general public can oversee the relevance of the issues at stake.

Discussion

Summarising the three worries that I have about the Dutch National Research Agenda, my view is that scientists are better at formulating questions and better able to see which solutions need to be formulated than anyone else. Moreover, from the perspective of legitimation it is doubtful whether science finds more legitimacy if the public can influence the scientific agenda.

I agree that more can be done to connect scientists with other partner organisations; and that scientists may need to be challenged to step into society to see what they can contribute. Choosing a direction of influence 'from society to research' has resulted in a set of questions that vary strongly with regard to the research agenda that has emerged from them. It truly concerns 'questions for science' rather than 'scientific questions', and I would have liked to see it the other around.

It should further be noted that already today a strong attachment between science and society is propagated in various ways. Through the 'top sector' approach, appointed fields receive extra research funding from the state in which businesses and scientists work together. This approach is not considered a success story (Koier et al., 2015; OECD, 2014). Likewise, in education research we see a heavy involvement of societal stakeholders in the agenda-setting of educational research through the Netherlands Initiative for Education Research (Nationaal Regieorgaan Onderwijsonderzoek, NRO).

Both the field (school organisations) and policymakers are represented in all layers of the NRO, including the fundamental research branch. Of note is that the NRO also involves the field in generating research questions, in ways similar to the NWA. And here too it would have been preferable if scientists had been stimulated to cooperate with partner organisations to help them formulate relevant research questions. It takes scientists to do the asking.

References

Achterberg, Peter, 'Een eenzame visserman op zoek naar geluk,' *Sociologie* 11, 2015, 51-78

Achterberg, Peter, Willem de Koster, and Jeroen van der Waal, 'A science confidence gap: Education, trust in scientific methods, and trust in scientific institutions in the United States, 2015,' *Public Understanding of Science*, 2015, 1-17

Bol, Thijs, Jacqueline Witschge, Herman G. Van de Werfhorst, and Jaap Dronkers, 'Curricular Tracking and Central Examinations: Counterbalancing the Impact of Social Background on Student Achievement in 36 Countries', *Social Forces* 92(4), 2014, pp. 1545-72

Dekker, Paul and Josje De Ridder, *Burgerperspectieven 2015* (The Hague: Sociaal en Cultureel Planbureau, 2015)

Green, Andy, John Preston, and Jan Germen Janmaat, *Education, Equality and Social Cohesion: A Comparative Analysis* (Basingstoke: Palgrave Macmillan, 2006)

Koier, Elizabeth, Barend van der Meulen, Edwin Horlings, and Rosalie Belder, *De Ontwikkeling van Vakgebieden in Nederland De Effecten van Beleid Op Het Nederlandse Onderzoeksprofiel* (The Hague: Rathenau Instituut, 2015)

Mazzucato, Mariana, *The Entrepreneurial State: debunking public vs. private sector myths* (London and New York: Anthem Press, 2013)

OECD, *Reviews of Innovation Policy: Netherlands 2014* (Paris: OECD, 2014)

Ultee, Wout, Wil Arts, and Henk Flap,. *Sociologie: Vragen, Uitspraken, Bevindingen*, 2nd ed. (Groningen: Wolters-Noordhoff, 1996)

Van de Werfhorst, Herman, Louise Elffers, and Sjoerd Karsten, *Onderwijsstelsels vergeleken: leren, werken en burgerschap* (Amsterdam: Didactief onderzoek, 2015)

Wilterdink, Nico, and Bart van Heerikhuizen, *Samenlevingen. Inleiding in de sociologie* (7th ed.) (Groningen: Noordhoff, 2013)

Skip the Agenda Building

Let the Wisdom of the Crowd Drive a Dynamic Tapestry of Science

Marten Scheffer[1] *and Johan Bollen*[2]

The Netherlands has recently conducted a broad popular survey in which the public were invited to submit online suggestions for the research questions and themes that they deem important. We applaud the idea of letting the public participate in a societal reflection on research priorities. The greater the number of participants and the broader their representation, the smaller the odds of missing relevant and important research areas. It helps science escape the trap of the ivory tower and reduces the risk of scientific tunnel vision. We therefore embrace the notion of stimulating a dialogue between the scientific community and the public. At the same time, we are wary of directing the public's energy towards the subsequent definition of a national science agenda to prioritize research themes. Science agendas that prioritize particular research areas are inevitably susceptible to bias and do not mitigate the widely perceived issues of how we presently prioritize and fund research. In our view this is a missed opportunity to really leverage the 'wisdom of the crowd' and make necessary improvements towards a more efficient, transparent, and equitable science funding system.

Problems of working with research agendas and peer-reviewed proposals

The present science funding system is based on painstakingly reviewing grant proposals, taking into account a variety of prioritized research themes and objectives. Although this system of strategic research agendas and peer-reviewed proposals has served us well, it now suffers from a number of broadly perceived concerns with respect to its ability to cope with the demands and scale of 21st-century science.

1. *Large overhead:* Scientists spend a disproportionally large part of their time writing and reviewing grant proposals, with very low odds of

---

1 Environmental Science Department, Wageningen University, the Netherlands.
2 School of Informatics and Computing, Indiana University, IN, USA.

actually receiving research funding. In addition, much time is spent on discussions about the prioritization of research themes. A large part of the available resources is thus lost in the process of allocating funding .

2. *Subjectivity:* Ranking and evaluating many excellent proposals easily devolves into an exercise in finding distinctions without a difference. This is demonstrated by the lack of correlation between the rankings produced by the evaluation of research proposals and the impact of the resulting work (Fang et al. 2016). We might do just as well by a random drawing of proposals, a procedure that would be equally fair and certainly more efficient.

3. *Excessive inequality:* A large fraction of available research funding ends up with a small group of scientists. This frustrates the scientific community, but it is also suboptimal with respect to a social cost-benefit analysis. We are not making good use of the available diversity of research talent, possibly amplifying cultural bias towards a select set of overrepresented groups.

4. *Artificiality:* The present system of science funding negates and ignores the important role of serendipity and flexibility that characterizes high-quality, innovative science. Most scientists accept the restrictions of the current project-focused system and its necessity of submitting multi-year plans in advance by deriving proposals from research that they have already conducted, but haven't yet published. This might be a good strategy to obtain research funding, but does not encourage innovation and serendipitous discovery.

Of these four issues, the first is perhaps the most pressing one. An exact determination of the current cost of the system remains difficult. However, recent estimates reveal that in Australia alone researchers spent more than five centuries' worth of research time on the submission of grant proposals (Herbert et al. 2013). These estimates do not include the time spent evaluating proposals, managing projects, writing project reports, defining and stipulating national research priorities, and the many other external costs of our grant peer-review system. Assuming that all these facets of the present proposal-driven funding machinery amount to 10-20% of researchers' time across universities, academic hospitals, and other institutes, we arrive at approximately 0,5-1 billion euros per year in the Netherlands (10-20% of the Ministry of Education, Culture and Science's budget for these institutes). Another rough calculation comes from Canada, where analysis of Natural Sciences and Engineering Research Council Canada (NSERC) statistics revealed that the $40,000 (Canadian) cost of preparing a grant application and having it rejected exceed that of giving every qualified investigator a direct baseline discovery grant of $30,000 (Gordon and Poulin 2009). We acknowledge that investing time in prioritizing research themes as well as writing and reviewing proposals might have inherent benefits. But do these outweigh the astronomical costs associated with the present system? If the present approach would result in something close to an optimum allocation of the funds that maximizes scientific innovation perhaps it would be worth it. But the strikingly poor correlations between review rankings and the impact of the resulting work (Fang et al. 2016)as well as high inequality in the distribution of funding suggest that this is not the case. The present system most likely does not effectively minimize costs and maximize scientific innovation. In fact, we might perhaps do better by simply skipping the entire procedure and awarding every applicant an equal and unconditional amount of funding (Gordon and Poulin 2009). We clearly need a careful examination of the return on investment of the present science funding system versus that of other possible systems.

Wisdom of the crowd as an alternative

In the remainder of this essay we ponder the possibility of distributing funds in a manner that wastes less money, but still acknowledges the different needs and productivity of individual scientists, avoiding the distortions resulting from the present funding machinery. The basic idea is that instead of evaluating and funding grant proposals, we distribute funding by evaluating the scientists themselves. Of course, this begs the question how this can be done in a reasonable, fair, and efficient manner. One possibility is to leverage the wisdom of the scientific crowd by involving all scientists, collectively, in the distribution of research funding to their peers. All scientists determine whom to best direct research funding to by making individual funding decisions with respect to their peers. The basic procedure to implement such a funding system can be simple and transparent (Bollen et al. 2014):

1. Every qualified scientist receives an equal and unconditional portion of the totality of available research funding.
2. Everybody anonymously donates 50% of the funding they receive to other, non-affiliated scientists, through a well-designed and easy-to-use website possibly managed by the national funding agency.
3. Repeat (1) and (2) so that those who receive a lot of funding must also distribute a lot of funding.

As funding circulates from one scientist to another, it settles into a fair distribution that respects the views and preferences of all scientists combined, without the requirement of submitting proposals, peer-reviewing them, managing projects, writing performance reports, defining research themes and mandates, etc. We should stress that there exists interesting mathematical work that underpins the effectiveness and efficiency of this system, which is why similar approaches are very common in other areas of the economy.

Of course, implementation of a workable and reliable version of this basic scheme requires careful elaboration of a number of aspects. First of all, we would have to decide who can participate in this system. As a first approximation, it could involve everyone with an academic position at an accredited institution. Secondly, it is of vital importance that conflicts of interest are prevented, e.g. by blocking donations to collaborators, co-authors, and individuals in the same institution. The system should be geared to detect the circulation of funding among small groups of colluding scientists. These measures would be similar to the rules that already apply in the present funding system, but one can imagine that a well-designed automated approach using detailed donation data may more effectively eliminate such problems. For instance, co-authorship and shared affiliations can simply be detected, and the same is true for collusion through reciprocal donations. The website where the participants select the names of scientists towards whom they direct the mandatory portion of their funds can show a stop sign upon detection of possible conflicts of interest and ask the participant to choose a different allocation.

Beyond the simplest scheme

This simple scheme can be extended in a number of ways. For instance, the redistribution percentage in the second iteration can be varied to result in either more equal or more 'merit-based' funding distributions. Simulations suggest that a 50% redistribution results in an inequality that roughly resembles the current skewness in the North American system (Bollen et al. 2014), whereas it is easy to see that an obligation to redistribute, say, only 5% in the second iteration round will result in a highly egalitarian distribution as most people receive only their equal minimum share. One can imagine that we could decide on an optimal level of inequality through the wisdom of the crowd as well, by asking participants what they consider a desirable difference between the richest and poorest in terms of received funding.

Another add-on that might be useful is to provide 'default' distribution options, e.g. 'redistribute my percentage equally to all scientists' or 'redistribute to all female environmental scientists'. Importantly, measurable bias (such as detected gender bias) can be corrected, for instance by raising the funding to each female scientist by a fixed percentage to achieve an unbiased male-female balance. This approach could also be applied to account for intrinsic differences in research costs between domains. For instance, experimental physics tends to be more expensive than theoretical physics. This brings us to another issue that requires some thought. Some lines of research or infrastructural projects need stable funding over multiple years. Sticking with the wisdom of the crowd as a leading principle, one option would be to offer the option of committing one's allocation for multiple years to the same group of researchers who have stated an interest in putting their funds together for such a project. Another possibility is to allow researchers to put up large common projects for funding. Whether such 'super-nodes' would indeed receive funding would remain up to the wisdom of the crowd. This might well make it more difficult to create powerful mega-projects. On the other hand, we have recently seen dramatic failures of seemingly attractive scientific megaprojects that illustrate the risk of making top-down decisions about where to direct public funds (Enserink and Kupferschmidt 2014, Fang et al. 2016, Margottini 2016). The wisdom of the crowd, since it is based on all available information in the system, could perform better at balancing the risks and rewards associated with such efforts.

Keeping the allocation of research funding firmly in the hands of the community reduces the distorting effects of lobbying, while saving a tremendous amount of time and money. Of course, it is possible to expand the definition of 'community' beyond scientists to allow the public, policymakers, and industry to be involved in the distribution weighting. For instance, one could decide to let 10% of the funds be distributed by 'the public vote'. This would stimulate public involvement and interest in the rich tapestry of our national research efforts without heavy-handed, top-down research agendas. Public influence would be accounted for in a transparent and efficient manner. Although it is crucial that the entire procedure remains transparent to the participants as well as the public, the anonymity of donors is paramount to ensure the system's effectiveness.

Unforeseen risks, benefits, and implementation

Self-Organized Fund Allocation (SOFA) addresses all four issues mentioned at the start of this essay, but it may also bring about fundamental changes in scientific communication. For instance, researchers will be incentivized to clearly communicate their plans and their work to the public and their peers, since this will stimulate donations. This reduces the 'ivory tower' effect and makes the scientific enterprise more open, transparent, and collaborative. On the other hand, it may carry the risk that funding will favour those that better promote their work and themselves. Again, the collective wisdom of the crowd may mitigate this issue. If many scientists see this pattern, they might very well decide to fund less visible, silent thinkers that actually need the funding.

Still, it remains impossible to foresee all the consequences, including psychological and social implications. Studies reveal that inordinate inequality leads to displeasure, whereas giving and participating leads to greater levels of satisfaction. SOFA could in this regard bring about positive changes for many researchers. On the other hand, presently well-funded researchers might risk a reduction of their research funding as a result of SOFA. Also, policymakers and administrators involved with the administration, management, and definition of national research priorities might see a sharp reduction in their workload and responsibilities. This raises the important question of whether the introduction of a SOFA-based funding system will be applauded by these constituencies. Obviously, we need to carefully consider these complex social and psychological consequences in designing an implementation process.

Moving to this system of Self-Organized Fund Allocation may seem like a leap of faith. We know the weaknesses of the current system, but how do we know if SOFA would do better? We can only really know it if we try it out. This does not have to happen at full scale immediately. In the Netherlands the allocation of all flexible research money amounts roughly to a yearly base of approximately 30,000 euros per researcher. However, one could run a trial with say 10% of the national research budget. If only active participants in the reallocation trial would receive their share of funding, the average gains of 3,000 euros per researcher should create enough incentive to participate. A multidisciplinary team can then take care of a repeated cycle of careful evaluation followed by adjustments to gradually improve the system over time, before scaling it up.

Between our writing and the moment that this essay went to press, the topic has made it into prime-time news, and the Dutch parliament has requested such an experiment.

Acknowledgements

We thank Ed Brinksma, Barend van der Meulen, and Marli Huijer for insightful discussions and comments on the ideas we present here.

References

Bollen, J., D. Crandall, D. Junk, Y. Ding, and K. Börner, 'From funding agencies to scientific agency', *Collective allocation of science funding as an alternative to peer review*, 2014, doi:10.1002/embr.201338068

Enserink, M., and K.J. Kupferschmidt, *Updated: European neuroscientists revolt against the E.U.'s Human Brain Project*, 2014, www.sciencemag.org/news/2014/07/updated-european-neuroscientists-revolt-against-eus-human-brain-project

Fang, F.C., A. Bowen, and A. Casadevall, 'NIH peer review percentile scores are poorly predictive of grant productivity,' *eLife 5*, 2016, e13323, doi:10.7554/eLife.13323

Gordon, R., and B.J. Poulin, 'Cost of the NSERC science grant peer review system exceeds the cost of giving every qualified researcher a baseline grant,' *Accountability in Research 16*, 2009, pp. 13-40, doi:10.1080/08989620802689821

Herbert, D.L., A.G. Barnett, and N. Graves, 'Funding: Australia's grant system wastes time,' *Nature 495*, 2013, 314, doi:10.1038/495314d

Margottini, L. *Why many Italian scientists aren't happy with a new, €1.5 billion research hub* (2016), www.sciencemag.org/news/2016/03/italys-plans-new-research-hub-get-critical-reception

An Economic Perspective on the Dutch National Research Agenda

Roel van Elk and Bas ter Weel

Introduction

The Dutch National Research Agenda consolidates a number of themes and routes that intend to help focus the scientific community on a number of core themes in the coming years. This implies that the research priorities are set with the objective of focusing and channelling research effort on what are perceived to be important scientific questions, societal challenges, and economic opportunities. The Dutch National Research Agenda aims to foster a better collaboration across different institutes and scientific disciplines and to increase the likelihood to stay at the research frontier by concentrating world-class research on a limited number of themes. An important question is whether or not setting such priorities makes sense to achieve the goals of scientific excellence, societal impact, and economic development. This essay discusses, from an economic point of view, the possible effects of such an agenda for science, society, and the economy. We first review the theoretical advantages and disadvantages of routing research effort. Next, we describe a number of trends and their implications. Finally, we address the implementation of a research agenda, with specific attention to the appropriate level of coordination and to its organisation.

Advantages of having a national research agenda

There are several theoretical arguments for building a national research agenda and routing scientific research into a number of themes. These arguments are mostly related to what economists refer to as market failures. These failures arise when engaging in research activities.

Economies of scale

The Dutch National Research Agenda aims to focus research activities on a limited number of scientific themes. This way of concentrating research effort is possibly valuable if there are economies of scale related to the production of knowledge. First, scale can be important for research activities because of fixed costs. Researchers often require expensive equipment, such as public labs, telescopes, or wind tunnels. The 2025 Vision for Science, which documents the government's ambitions with respect to science policy, has announced the establishment of a permanent committee responsible for the coordination of investments in large-scale research infrastructure (Ministry of Education, Culture and Science, 2014). Research infrastructure is of interest both for conducting basic and applied research. In a recent letter to Parliament (No. 2016Z04755/2016D10344), the Dutch Minister of Economic Affairs addressed the introduction of a specific strategic agenda for applied research facilities.

Second, scale can be important because knowledge spillovers are crucial. Concentrating research effort on specific themes can foster scientific production because of an increased exchange of knowledge and creative ideas.

Contributing to the progress of science is complex and requires a team of complementary workers who each contribute with their specific skill and knowledge. A sufficient number of researchers is needed for gaining from such patterns of specialization or to allow interdisciplinary work, while fragmentation of research activities leads to suboptimal outcomes. Setting research priorities may help create a sufficient mass per theme to benefit from this complementarity. This increases welfare if the 'market' for research does not reach the optimal level of concentration. The 'market' refers both to the private sector (with the objective of profit maximization) and the scientific community (with the objective of producing knowledge).

It is not immediately clear why the market would not reach an efficient scale and why the government would do better by setting research priorities. A lack of critical mass in universities may result from the fact that they have been operating within national boundaries and national institutions that limit incentives for performance. This may cause scattering of research activities and underutilization of complementarities in research.

The importance of scale likely differs across research disciplines. For example, biomedical sciences require on average more costly research infrastructure than social sciences. Expenditures on research equipment are estimated to cover around 15-25 percent of total research budgets in capital-intensive disciplines (e.g. biomedical sciences, physics, and engineering), and around 5-10 percent in other disciplines (Rathenau Instituut, 2009, p. 46/47). Developments in the availability of more data and new techniques to utilize and store these data are also likely to increase fixed costs in social sciences.

Information problems

A second type of market failure that could legitimize a centralized routing of research effort is incomplete information. This refers both to information problems with respect to the most valuable research activities and to coordination problems among potential research collaborators.

Information on the most promising research activities

Directing research effort by the government is likely to be beneficial if the government has a better view on the most important or promising research areas. Yet the government faces the same information problems as the market, making picking the set of most promising projects an extremely difficult task. Fundamental research is inherently uncertain and, if anything, one would expect researchers to be better informed than the government. This also relates to the involvement of citizens, who in addition are likely to be less well-informed than researchers. An advantage of bringing together the preferences of scientists, citizens, firms, and the government could be that information is shared which could help to create a social basis for investing in science. In addition, principal agent problems could be mitigated.

Information problems and directing research efforts are closely related to the way public research funds are allocated. In the Netherlands, around 70 percent of the public funds are allocated based on institutional funding, and around 30 percent of the public funds are allocated in competition to pre-screened research projects. The latter type of funding helps to solve information problems. The screening of research proposals increases the likelihood that resources are devoted to the most promising projects (assuming that quality differences across proposals are well observable). This type of funding is also well-suited for directing resources to specific groups of researchers or research areas. Institutional funding, after all, implies that the government leaves control to universities or public research institutes concerning the allocation of funds to fields of research. A disadvantage of project-based competitive funding is that the screening process can be costly because of the required time for judging and writing (non-granted) research proposals. In addition, it may have adverse consequences for investments in risky, long-term research activities (e.g. Manso, 2011; Azoulay et al., 2011). A single best funding type does not exist. Empirically, there does not seem to be a clear relationship between a country's share of project-based competitive funding and its research performance in terms of publications or citations (Van Dalen et al., 2015, p. 10).

Coordination of research activities
Another potential reason to direct scientific research investments would be if coordination problems lead to insufficient collaboration. First, public and private research institutes may have conflicting goals that hamper combined research initiatives. For example, researchers at public institutes aim to publish new research fast (the standard of disclosure) because publications are important for their reputation and career perspectives. This fosters transparency and openness of research. Private research institutes, however, are more likely to keep new knowledge to themselves, at least until intellectual property rights have been acquired or profitable products have been launched in the market. These conflicting incentives could hamper successful collaboration and the valorisation of basic research. The enhancement of public–private collaboration is one of the main purposes of the Dutch top-sector policy that was launched in 2011. Currently amounting to a total investment of around 1 billion euros, this policy consists of several subsidy and organisational measures targeted at pre-selected sectors that have been labelled crucial to the Dutch economy. Among the identified sectors are high-tech systems and materials, life sciences and health, and the agro and food sector. By aligning the goals of private firms and public research institutes the policy has the potential to stimulate collaboration and the diffusion of knowledge. A potential drawback of earmarking resources for specific sectors, however, is that it is likely less focused on basic research and long-term research goals. Building on areas that have been successful in the past brings about the risk of conservatism. An additional risk is that it could hamper research on general purpose technologies. Such technologies might not be especially important from the perspective of a single sector, but could be of great importance for long-term economic development.

Second, research institutes can choose their own priorities, without taking into account the priorities or goals of the other institutes. This may lead to dispersion of resources and activities ('stepping on toes'). Independent priority setting by actors in the Netherlands, such as universities, the Netherlands Organisation for Scientific Research (NWO), and the central government, does not seem to have led to a set of clear research priorities at the national level (Rathenau Instituut, 2010, p. 59). Priority setting by the government may help coordinate research activities and reduce dispersion.

Third, coordination by the government could foster interdisciplinary research. Spillovers across different areas of specialization can be particularly valuable for challenging, fundamental research topics, for exploring new fields of research, or for solving social problems.

Externalities

Some research comes with larger externalities than other activities. For example, research on mitigating the effects of climate change will likely have positive spillovers for many people and for future generations, whereas other research output has smaller spillovers. In case of large differences in spillovers across research themes and social problems, funding these themes can help to internalize positive spillovers to the benefit of society at large.

Scientific researchers are not always likely to take up research topics with the largest externalities. First (more relevant for the private sector), large externalities imply that individual researchers or research groups can only reap a relatively small part of the benefits of their research efforts. Therefore, private firms have relatively low incentives to focus on social challenges that do not foster profits. For example, innovative clean technologies can yield benefits in terms of a better protection of the environment, which are not taken into account by individual firms. Second (more relevant for the public sector), publication incentives affect the research agenda. A long list of publications yields reputation and career perspectives. This encourages the dissemination of knowledge, but may hamper research that benefits society at large. 'Publish or perish' implies that researchers choose topics that most likely will result in publication in academic journals. Those articles do not necessarily deal with topics in which science can contribute most to solving social problems. The government could help directing research to solving social challenges that are not brought about by the market. Such a strategy by the government is, however, not completely straightforward. Short-sightedness and (potentially conflicting) interests of politicians could lead to socially suboptimal choices.

The entrepreneurial government

Next to correcting market failures, it has been argued that the government should have a more prominent role in the innovation system. Through the big bets it makes on new technologies it creates and shapes the markets of the future and can help solve social problems. In the United States, for example, the government has played an important role in realising breakthroughs in areas such as space research, biopharma, and the internet (Mazzucato, 2013). Specific government-funded projects and collaboration between scientists and entrepreneurs have led to substantial economic payoffs in the private sector and to new opportunities for society. It is not a priori clear, however, what the outcomes would have been in case of a different use of public resources because there is no counterfactual policy.

Disadvantages of directing scientific research

Disadvantages of directing scientific research are mostly related to government failures due to information problems and to the negative consequences of a low level of flexibility and diversification. These could facilitate a suboptimal allocation of resources across research fields or projects. In addition, setting strong priorities by the government could undermine the attractiveness of the Netherlands for scientific talent.

Government failure

It is difficult for the government to determine the social returns of specific research topics or projects. If anything, researchers are more likely to be well-informed about the most promising and practicable research projects. Given the information problem, it seems sensible to involve researchers and firms in the process of priority setting. Still, this does not guarantee optimal choices. Researchers and users may favour 'hot topics' which have received a lot of attention recently (for example because of recent breakthroughs) or which have the greatest chance of getting published in top-ranked academic journals. This may lead to hypes but also to conservatism if most of the resources are devoted to current strengths and not to long-term research goals. In addition, firms' focus can be on especially commercially interesting topics, or topics that appeal to the imagination, such as technological breakthroughs at the expense of a knowledge base about foreign languages to fight terrorism. It is difficult for the government to recognize such kinds of strategic behaviour and to maintain a broad portfolio of research areas (within the limits of the budget). In addition, the process of information gathering is costly and may have unintended effects, such as lobbying and rent-seeking behaviour. Moreover, apart from the theoretical optimal choices, it may be difficult to realise an optimal allocation in practice due to agency problems. The government seems to be unable to completely control activities and incentives of universities and researchers.

Low level of flexibility and diversification

Resources that are devoted to specific topics are not easily transferred to other topics. Hence, dynamic adjustments to new information or actual developments are difficult to establish. This could be an important drawback since it is not straightforward that current strengths are permanent strengths. A policy of diversification has the advantage of flexibility. This also allows for small-scale experiments in different fields to obtain more insights in the perspective of future research and investments. Consequently, targeted additional resources can be devoted to those topics that have shown to be most promising. In this way effective selection processes could contribute to better research choices.

Inflexibility is strengthened if influential researchers or politicians have special interests in a continuing focus on particular research themes. Researchers are likely to continue their own research programme or extend it with new elements. This can lead to 'overshooting' if it prevents resources from being transferred to more promising and new research areas. In addition, extending specific topics may lead to lower quality because researchers are scarce. If the availability of researchers with relevant expertise in a single research topic is limited, additional resources are likely provided to less productive researchers.

An additional risk of too little diversification is that it undermines the general knowledge base needed for absorbing knowledge from abroad. Striving for excellence in specific fields may come at the expense of building knowledge in other fields. A sufficient level of knowledge in those latter fields, however, is still needed to be able to use research produced by others.

Adverse effects on attracting or binding talent

Attracting and binding scientific talent is an important element of science policy in the Netherlands (Ministry of Education, Culture and Science, 2014). Dutch science seems to be quite attractive for foreign researchers. Dutch universities are placed relatively high in worldwide university rankings, such as the Shanghai Ranking. Universities are internationally oriented, and English serves as a *lingua franca* in educational and research programmes. In addition, a PhD track in the Netherlands is attractive because of the position of the PhD student as an employee. In a globalizing research market with increasing international competition, the Dutch government aims to be a continuing breeding ground for talent. Setting strong research priorities, however, could reduce the attractiveness of research positions. Researchers may be less inclined to come to (or stay in) the Netherlands if they are not autonomous in setting their own research agenda. Empirical evidence has shown that researchers value academic freedom highly. Scientists seem to be willing to pay for being allowed to pursue and publish an individual research agenda (Stern, 2004, p. 835). Hence, limited opportunities to set up an own research agenda could lower the attractiveness of an academic career in the Netherlands.

Developments in the market for science

There are economic reasons for directing scientific research. At the same time, directing research efforts has several drawbacks. It is not a priori clear whether the advantages outweigh the disadvantages. However, recent developments in the market for scientific research, such as rapid knowledge accumulation, increased internationalization, specialization, and teamwork, seem to make the case for concentration of research activities more plausible.

The worldwide scientific output has increased rapidly over time. Since the 1960s, the annual growth rate in publications has averaged 5.5 percent (Jones, 2011, p. 104/105). This implies that the annual number of journal articles published has doubled every 13 years. Because the total stock of knowledge is strongly accumulating, researchers naturally respond by narrowing their area of expertise. This may help to explain the importance of teamwork in academia (e.g. Black and Stephan, 2008). Increasingly teams, instead of individuals, generate scientific contributions. Mean team size had risen at rates of 15-20 percent between 1960 and 2010. The shift towards teamwork has been observed in almost all subfields of research (e.g. Wuchty et al., 2007; Jones, 2011). In science and engineering mean team size increased from approximately 3.1 in 1990 to 4.2 in 2005, compared to an increase from around 1.6 to 2.1 in the social sciences. There is also empirical evidence that collaborative efforts produce higher-quality research output. Team-authored papers published between 1995 and 2005 received more than twice as many citations as single-authored papers. This holds for science and engineering as well as the social sciences (e.g. Wuchty et al., 2007; Jones, 2011).

The market for scientific research has become increasingly globalized. ICT developments have fostered the international flow of ideas. The European Research Area (ERA), established in 2000 with the aim of creating a unified research area across Europe, has created a single market for scientific research. The unification of higher education degrees after the Bologna declaration in 1999 has fostered the international mobility of researchers within Europe (Curaj et al., 2012). In addition, many universities in Europe and Asia have experienced various reforms during the last decades, which enabled them to become important players in the global higher education market (Clotfelter, 2010, p. 12/13). The internationalization of PhD positions is a worldwide trend. In highly developed OECD countries, the average share of foreign PhD students has increased from 16 percent in 2006 to 23 percent in 2012. In the Netherlands, the share of foreign PhDs is relatively large, around 40 percent. The total number of foreign PhD candidates employed by Dutch universities increased from around 2,300 to almost 4,000 between 2005 and 2013 (Van Elk et al., 2016, p. 5).

These developments have led to an increased competition for funding and talent and have also stimulated specialization of research activities. Specialization helps to create excellence because it allows exploiting comparative advantages in specific research areas and a better allocation of researchers across institutes. If researchers with a particular specialization work together, various types of knowledge and ideas are likely to be exchanged and used in the creative and innovative process. International collaboration has increased in recent decades and the higher average citation impact of team publications is typically even larger when co-authorship is taking place within an international team of researchers (Adams, 2013, p. 559).

Specialization and the tendency of increasing scale imply that a fewer number of research topics, and hence choices for particular research fields, can be addressed (by a fixed number of researchers and a given budget). Especially for small countries, with relatively limited resources, concentration of research topics seems important to perform excellent research. In an international market, specialization also seems to be a less risky avenue because research crosses national borders easily. At the same time, the need for absorptive capacity for research from abroad is increasing. Focusing on particular research areas implies less diversity and fewer activities in other areas. While striving for world-class research in specific fields, it seems important to take into account potential consequences for the general knowledge base needed to understand and use research from abroad.

The implementation of a national research agenda

The practical implementation of a national research agenda relates to questions about the appropriate level at which research activities should be coordinated as well as some organisational issues, including the choices for particular research areas.

Level of coordination: national or supranational research agenda

An important question is whether coordination should take place at a national or a supranational level. Arguments for supranational (European) coordination are related to the identification of global research topics and to mitigating free-riding behaviour. There is an increased focus on global research themes that ask for international cooperation, such as climate change, demographic changes, or the transition of clean energy. This suggests that supranational coordination is beneficial, since research agendas at national levels could still conflict and lead to dispersion or inefficient use of resources at the higher level. In addition, if scientific knowledge has the characteristics of a public good (non-rivalry and non-excludability), country A can benefit from knowledge produced by country B, and vice versa. This may lead to 'free-riding' by national governments and a decrease in global investments in science. Supranational coordination of research activities is then needed to realise the socially optimal investment levels. Developments in ICT increase accessibility to codified knowledge, which could increase the use of scientific knowledge produced by other countries, and hence the need for supranational coordination.

On the other hand, there are several arguments for national coordination of research themes. First, despite ICT developments distance still matters in the diffusion of knowledge. Whereas codified knowledge can be exchanged relatively easily (for example through the internet), tacit knowledge requires personal contact. Hence, free-riding on research from abroad is not straightforward and geographic proximity can be helpful or even necessary in capturing the benefits from knowledge spillovers (e.g. Audretsch and Feldman, 1996; Belenzon and Schankerman, 2013). Second, country-specific challenges may require country-specific research investments. For example, research on water safety could be of special importance for the Netherlands. National research investments can be used to solve country-specific problems rather than global challenges. Finally, and more generally, the development of the knowledge economy may encourage setting national science priorities. Knowledge has become increasingly important for productivity growth. It is thus of crucial importance for countries to be capable of developing new technologies, and/or understanding and absorbing scientific or technological developments in other countries.

Organisation of a national research agenda

Several choices can be made with respect to the implementation of a national research agenda. An important choice is whether or not to actively cooperate in international frontier research or to focus on specific national challenges, such as for example water safety. In the latter case a country can benefit from research performed by other countries (free-riding), whereas investments in science are specifically targeted towards national topics. This case obviously also requires investments in education and science to ensure sufficient 'absorptive capacity' to be able to use new scientific insights produced by others. The advantage of the first case is that it contributes to access to international scientific networks and links with the international science base, and it fosters cross-country collaboration. This can also result in additional research funding from abroad. In this respect it is noticeable that the European Union is likely to become an increasingly important player in research activities. At this level it is easier to create efficient and sufficient mass, competition, and specialization, which is further stimulated by the steady increase of European research funding in recent years (up to 80 billion euros in Horizon 2020).

Finally, two remarks seem in place when it comes to implementing a national research agenda. First, it seems functional to ensure that, next to targeted research activities, there remains sufficient potential for open and fundamental research. This type of research is intrinsically valuable, may attract researchers, and has the potential of substantial long-term contributions. Second, even after the implementation of a national research agenda, it remains important to learn more about optimal ways of spending research budgets. In this respect it is valuable to monitor the research agenda, and – more generally – to invest in evaluations of specific institutions or science policy measures.

References

Adams, J., 'The Fourth Age of Research', *Nature,* 497, 2013, pp. 557-560

Audretsch, D., and M. Feldman, 'R&D spillovers and the geography of innovation and production', *American Economic Review,* 86(3), 1996, pp. 630-640

Azoulay, P., J. Graff Zivin, and G. Manso, 'Incentives and creativity: Evidence from the academic life sciences', *RAND Journal of Economics,* 42(3), 2011, pp. 527-554

Belenzon, S., and M. Schankerman, 'Spreading the word: Geography, policy, and knowledge spillovers', *The Review of Economics and Statistics,* 95(3), 2013, pp. 884-903

Black, G., and P. Stephan, 'The economics of university science and the role of foreign graduate students and postdoctoral scholars', in *American Universities in a Global Market,* edited by C. Clotfelter (Chicago: University of Chicago Press, 2010), pp. 129-161

Clotfelter, C. (ed.), *American Universities in a Global Market* (Chicago: University of Chicago Press, 2010)

Curaj, A., P. Scott, L. Vlasceanu, and L. Wilson (eds), *European higher education at the crossroads: between the Bologna process and national reforms* (Dordrecht: Springer Science & Business Media, 2012)

Jones, B., 'As Science Evolves, How Can Science Policy?', in *Innovation Policy and the Economy*, edited by J. Lerner and S. Stern (Chicago: University of Chicago Press, 2011), pp. 103-131

Manso, G., 'Motivating innovation', *Journal of Finance*, 66(5), 2011, pp. 1823-1860

Mazzucato, M., *The entrepreneurial state: debunking public vs. private sector myths* (London and New York: Anthem Press, 2013)

Ministry of Education, Culture and Science, *2025 Vision for Science: choices for the future* (The Hague: Ministry of Education, Culture and Science, 2014)

Rathenau Instituut, *Investing in Research Facilities* (The Hague: Rathenau Instituut, 2009)

Rathenau Instituut, *Focus and Mass in Scientific Research: The Dutch Research Portfolio in an International Perspective* (The Hague: Rathenau Instituut, 2010)

Stern, S., 'Do scientists pay to be scientists?', *Management Science*, 50(6), 2004, pp. 835-853

Van Dalen, R., R. van Elk, and D. van Vuuren, *Public Research Funding: The Pros and Cons of Different Funding Methods*, CPB Policy Brief 2015/07

Van Elk, R., I. Rud, and B. Wouterse, *The Economic Effects of Foreign PhDs*, CPB Policy Brief 2016/01

Wuchty, S., B. Jones, and B. Uzzi, 'The Increasing Dominance of Teams in the Production of Knowledge', *Science*, 316, 2007, pp. 1036-1039

What is the Good of Government Interference in Science?

A Question from Late Nineteenth-Century Germany

Herman Paul

Abstract

What are the goals pursued by government interference in science? Drawing on a historical case study, this chapter examines whether anything can be learned from how the German physicist Carl August Voller (1842-1920) assessed such government interference in the decades around 1900. Although Voller's distinction between 'practical' and 'scientific' research is a typically nineteenth-century dichotomy, which as such is unlikely to be of much value in our current situation, Voller's question what are the 'goods' or 'aims' favoured by states or city councils committed to science funding or regulation may still serve as a starting point for analysing what is at stake in contemporary struggles with government interference in science.

Introduction

Through a quirky twist of fate, the German physicist Carl August Voller (1842-1920) has become best-known for the fingers of his right hand. This hand was among the first on which the future Nobel laureate Wilhelm Röntgen (1845-1923) tried out his X-ray techniques, which resulted in a photograph that was reproduced countless times after its first publication in a 1896 report on Röntgen's discovery (Glasser, 1959, pp. 26-27). Far less known than his bony hand is Voller's research on electricity and electromagnetism, his contributions to science education, most notably in a booklet on electromagnetic telegraphy, and his critical views on the politics of science in Wilhelmine Germany. This is not surprising: Voller only played a minor role in German physics around 1900. For the better part of his career, he was not even an academic. He received an honorary professorship only shortly before his death, in acknowledgment of his contributions to the foundation of the University of Hamburg in 1919. If he made an impact, it was on education, popular science, and political life in Hamburg more than on physical research in Wilhelmine Germany (Voller,

1873, p. 57*; Neubert 1905, p. 1513; Weinmeister 1925-1926, p. 1318). Nonetheless, precisely because Voller was a rank-and-file physicist, whose commitment to scientific research outweighed the support he received for it, his remarks on research in a government-funded context warrant a moment of attention.

As director of the Physical State Laboratory in Hamburg, Voller experienced how true the adage was that he who pays the piper calls the tune. On the one hand, the *Staatslabor* testified to the prestige that physical research enjoyed in late nineteenth-century Germany. Developed out of a 'physical cabinet' in the local gymnasium, the Physical State Laboratory was a well-equipped institute with possibilities for conducting serious research, especially after its move in 1898 into a spacious modern facility (Witte, 1985, pp. 9-10). On the other hand, the city magistrates who funded the lab often treated it as a scientific consultancy agency whose help they could enlist in analysing current problems such as lightning strike patterns and groundwater levels in the city on the Elbe (Voller, 1891a, p. lxxiv). Moreover, in addition to delivering public lectures on a weekly basis, the director was supposed to hold daily office hours for citizens and civil servants with physics-related questions on their minds (Voller, 1886, p. lxiv). With so much time and resources drawn away from independent research, it is unsurprising to find Voller emphasizing the need for his lab not to economize on 'serious scholarly work' (Voller, 1901, p. 208).

Moreover, in the long-standing controversy over the relative merits of 'pure' and 'applied' research (on which more below), Voller often emphasized the importance of the former, arguably because he spent most of his time doing applied research. This became particularly clear in 1891, when the death of Wilhelm Eduard Weber (1804-1891) unleashed a flood of necrologies trying to claim the famous Göttingen physicist as a forerunner of either 'pure' or 'applied' modes of research. Not a few obituaries honoured Weber, together with Carl Friedrich Gauß (1777-1855), as the inventor of the telegraph and, consequently, as a practically-minded and application-oriented scientist – even though the kilometres-long wire through which Weber had connected his laboratory to Gauß's observatory, back in 1833, had not allowed for transmitting more than six words per minute.[1] Voller, by contrast, rejected the idea that his much-revered teacher had conducted research with an eye to such practical applications as a telegraph. Weber's aim, wrote Voller, had not been to provide society with technical devices, but to further the cause of science. He had been driven by 'purely scientific goals', which were irreconcilable with what Voller condescendingly called 'the unrest and loud noise' of the 'market of contemporary life'. Consequently, in Voller's portrayal, Weber

---

1   E.g., 'Prof. Wilhelm Eduard Weber', *Naturwissenschaftliche Wochenschrift*, 6 (1891), p. 275.

emerged as a model of ascetic virtue, whose 'versatility of mind' flourished in a 'silence' and 'rest' of which Voller, busy in meeting the demands of his Hamburg superiors, could only dream (Voller, 1891b, pp. 29-30).

What shall we make of this physicist, eager to seize the opportunities that the city of Hamburg offered him (Voller, 1884, pp. lxix-lxx), ready to give advice on matters varying from clinical thermometers to electric illumination of the St Michael's church tower (Voller, 1894, pp. lxiii), but firm in his insistence that such 'practical' work did not deserve the name of 'scientific' research (Voller, 1891a, p. lxxi)? How convincing is his view that government interference in science contributes to a 'noise' disrupting the 'silence' required for pure academic work? In other words, does Voller, the late nineteenth-century physicist, have anything to say to us, early 21st-century scholars struggling with new models of research funding and government interference in science?

One possible way of making Voller's case fruitful for contemporary reflection consists of assessing as specifically as possible the strengths and weaknesses of Voller's position or by arguing, from a contemporary point of view, that although some of his arguments have become outdated, others are still pertinent to contemporary reflection. This, however, is not what I understand to be the historian's task. As a historian, I am more interested in reconstructing Voller's questions than in evaluating his answers. Historians enrich contemporary reflection not so much by bringing in historical precedents to current opinions ('Voller said it as early as a century ago'), but by confronting contemporary audiences with questions from the past that tend to be ignored or silenced in the present (Paul, 2015, pp. 123-138). Consequently, rather than focusing on Voller's distinction between 'practical' and 'scientific' research – a typically nineteenth-century dichotomy, which as such is unlikely to be of much value in our current situation – this chapter will focus on Voller's worries about science put in the service of non-scientific ends. It will argue that his question regarding the 'aims' favoured in such circumstances, or the 'goods' that are being pursued, may still serve as a starting point for analysing what is at stake in government interference in science, without necessarily making us as pessimistic as Voller about the 'unrest and loud noise' characteristic of modern scientific life.

'Pure' versus 'applied'

Whenever nineteenth-century scientists raised their voices to praise the 'scholarly attitude', 'shrewdness', or 'indefatigability' of their predecessors, as Voller did in his obituary of Weber (Voller, 1891, p. 29), it was likely that they did so with a polemical agenda in mind (Echterhölter, 2012). This was not just because virtues were regarded as corrective by nature: 'indefatigability' was a virtue only because human beings could be tired, weary, and lazy (Foot, 1978, pp. 8-9). More important was that virtues tended to be emphasized in contexts where they were perceived as lacking or absent. Defenders of 'impartiality' made themselves heard when they felt that the virtue in question did not enjoy the prestige it deserved, just as 'imaginativeness' was typically recommended as an antidote to a perceived rise of empiricist fact fetishism. Virtue language, in other words, emerged in contexts of reflection on the vocation of the scientist in relation to dominant trends in the field. Epistemic virtues in particular – virtues aimed at epistemic goods such as knowledge and understanding of reality – were highlighted at moments when such virtues were seen as being threatened by non-epistemic motives, such as desires for money or fame (Paul, forthcoming; Saarloos, forthcoming).

Much the same is true for the nineteenth-century concepts of 'pure' and 'applied' science, on which Voller obviously drew in distinguishing between 'practical' and 'scientific' research. When in 1883 the American physicist Henry A. Rowland (1848-1901) made his famous plea for 'pure science', he did so out of contempt for scholars whom he saw as forsaking their vocation by engaging in 'applied' consultancy work:

> There are also those who have every facility for the pursuit of science, who have an ample salary and every appliance for work, yet who devote themselves to commercial work, to testifying in courts of law, and to any other work to increase their present large income. Such men would be respectable if they gave up the name of professor, and took that of consulting chemists or physicists (Rowland, 1884, p. 110).

This in turn prompted Robert H. Thurston (1839-1903), professor of mechanical engineering, to defend the modern invention of 'applied science' against the 'old spirit [...] of reverence for the non-utilitarian element in science' – a spirit that 'still survives', as Thurston observed with an implicit nod to Rowland, but fortunately exercises only limited influence 'upon our modern life' (Thurston, 1885, p. 237).[2] Equally belligerent was the chemist Ira Remsen

---

2   Helpful analysis of this exchange is provided in Ronald Kline, 'Construing "Technology" as "Applied Science": Public Rhetoric of Scientists and Engineers in the United States, 1880-1945', *Isis*, 86 (1995), pp. 194-221 and Paul Lucier, *Scientists and Swindlers: Consulting on Coal and Oil in America, 1820-1890* (Baltimore, MD: Johns Hopkins University Press, 2008), pp. 313-323.

(1846-1927), a future colleague of Rowland at Johns Hopkins University, who recommended a firm dose of 'German thoroughness' to remedy the 'Americanization' of science, defined a few years later by *The Popular Science Monthly* under reference to Emil du Bois-Reymond (1818-1896) as 'the growth of the utilitarian spirit, which is gradually substituting immediate, practical, wealth-yielding studies for the more elevated, disinterested, and ennobling intellectual pursuits which have been cherished in past times' (Remsen, 1872, 33-34; Remsen, 1878, p. 495). In short, 'pure' and 'applied' were combative terms or battle cries invoked in situations of scholarly disagreement over research priorities or trends in academia.

In recent years, historians of science have done much to historicize those battle cries. Graeme Gooday, for instance, has argued that their semantic range was so wide that 'pure' and 'applied' cannot be treated as a simple dichotomy or as opposite ends of a linear scale. When in 1870 the British chemist Alexander W. Williamson (1824-1904) published *A Plea for Pure Science* – thirteen years before Rowland's essay with the same title – he did not define 'pure science' as non-utilitarian, as Rowland would do, but as a form of research committed to the same goals as its 'applied' counterpart. For Williamson, fundamental research was needed for realising even greater practical benefits in the long term than research focused on immediate results could possibly bring about. If this illustrates the unstable meanings associated with 'pure' and 'applied' research, Gooday also draws attention to the political subtexts of the discourse. What 'pure' and 'applied' meant depended in no small degree on the agendas of the speakers. Williamson's *Plea* was, among other things, a defence of the university as a privileged site of inquiry as well as a request for state patronage of academic research. So, whenever we find nineteenth-century scholars debating the relative merits of 'pure' and 'applied' science, we have to be attentive to the contested nature of those phrases and realise that they were combative terms used in the service of sometimes outspoken political agendas (Gooday, 2012, pp. 547-549).[3]

On top of this, historians have drawn attention to the philosophical and psychological baggage carried by such adjectives as 'pure' and 'applied'. Paul Lucier, most notably, has shown that Rowland and his critics assumed rather different views of human nature. While Rowland's ideal of pure science presupposed the sober view that human beings are typically unable to resist pecuniary temptations, advocates of 'applied science' tended to be more

---

3 See also Robert Bud, '"Applied Science": A Phrase in Search of a Meaning', *Isis*, 103 (2012), pp. 537-545.

optimistic about 'the ability of individuals to manage money and its allure' (Lucier, 2012, p. 536). This anthropological difference had epistemological implications to the extent that it made different demands on a scholar's motivation for scientific research. Whereas Thurston *cum suis* had few problems with mixed motives, Rowland and his likes emphasized that science had to be practised 'for its own sake', that is, without ulterior motives. 'Purity', then, was a 'purity of motive' in the first place (Kevles, 1979, p. 141). All this shows how context-dependent the meanings of 'pure' and 'applied science' were – and how unhistorical it would be to expect from any of the nineteenth-century scientists mentioned so far a ready-made answer to the question how we, more than a century later, can wisely navigate our own dilemmas of state-sponsored research and government interference in science.

Aims of science

Still, a historicizing of the terms in which German, British, and American scientists of Voller's generation tried to make sense of the challenges they faced is not the same as declaring the worries articulated in those terms as irrelevant to future generations. Even though the *positions* adopted by Voller, Rowland, Thurston, and Williamson may be too dependent on nineteenth-century contexts and assumptions to be helpful for present-day purposes, the *questions* to which they responded are not thereby irrelevant. Indeed, one does not have to subscribe to an idealist notion of time-transcendent issues – a *Sache*, as the German philosopher Hans-Georg Gadamer (1900-2002) famously called it – to acknowledge that scientists in the past sometimes struggled with questions that an imaginative mind can recognize as somehow related to questions that we are facing, or should be facing, in the present. Admittedly, scientists have devoted themselves to many questions that we find difficult to recognize as valid or important – think only of Isaac Newton (1642-1727) and his fascination for alchemy and Biblical chronology. So I am not making a sweeping claim about relevance over time of questions that occupied scientists in times different from ours. Some questions are simply more pertinent to present-day concerns than others. Relevance, moreover, does not require analogous circumstances: the past always differs in important respects from the present. It is precisely in this difference, though, that questions from the past may acquire a sometimes unexpected relevance. In their 'otherness' or 'foreignness', so to speak, they may raise a thought or offer a perspective that may be fruitful for present-day reflection (Paul, 2015, pp. 123-138).

To what extent is this the case for Voller's question, formulated in 1890s Hamburg? As we saw above, the director of the Physical State Laboratory used the concept of 'goals' (*Zwecken*) to distinguish between Weber's 'scientific' aspirations and the 'practical' work that occupied most of his own time. Rowland and Thurston likewise defined 'pure' and 'applied' research in terms of their 'aims', 'goals', or 'missions'. While Rowland called for commitment to developing 'a better and more glorious idea of this wonderful universe' (1884, p. 122), Thurston articulated a utilitarian Baconian commitment to improvement of the human lot:

> The mission of science is the promotion of the welfare, material, and spiritual, physical and intellectual, of the human race. [...] It is charged with the duty of seeking the cause of every ill to which mankind is subject; of finding a remedy for every misfortune to which the race is now liable; of ameliorating every misery known to sage or savage; of seeking ways of giving to all every comfort and all healthful luxuries... (Thurston, 1885, pp. 231-232)[4]

In its broadest possible phrasing, then, Voller's question concerned the 'aims of science', to use the phrase that philosophers of science have been employing since Stephen Toulmin (1922-2009) (Toulmin, 1961) and Larry Laudan (1984). What are the aims that scientists ought to pursue? What are the goods to which they ought to be committed? For Voller and his contemporaries, this was in part a question of personal motivation and hence of desire, will, and virtue. Voller highlighted Weber's virtuousness, not only because this was what nineteenth-century genre conventions required, but also, more importantly, because Weber's 'deeply serious scientific attitude' (*tief-erstem, wissenschaftlichen Sinn*) and 'purely scientific will' (*reinem wissenschaftlichen Wollen*) served as markers of his 'purely scientific goals' (Voller, 1891b, p. 29). Personal motivation, however, was not the only thing that mattered. As illustrated by Rowland's warning against monetary temptations, personal motivation could be influenced by outside factors: non-epistemic desires could be stirred with the effect of epistemic commitments being compromised. Consequently, the scientist's individual fight against temptation needed institutional support. Not only individuals, but institutions, too, had to commit themselves to proper aims of science. In

---

[4] On Thurston's Baconian commitments, see Kline, 'Construing Technology', p. 202 and Edwin T. Layton, Jr., 'American Ideologies of Science and Engineering', *Technology and Culture*, 17 (1976), 688-701 (p. 694).

Lucier's summary of Rowland's argument: 'Where men of science are weak, the walls of universities had to be strong' (Lucier, 2008, p. 318).

But how likely was it that scientific institutions, academic or otherwise, would be committed solely to epistemic aims? If Voller experienced anything, it was that the Hamburg city magistrates had other than 'purely scientific' goals. The Physical State Laboratory, moreover, was not the only scientific institute funded by sponsors whose aims differed in kind or degree from those that Voller preferred. As Jonathan Harwood has shown, the rivalry between universities and technical colleges in late nineteenth-century Germany as well as the tense relation between the Prussian Academy of Sciences and the Kaiser Wilhelm Society for the Advancement of Science were caused in great part by different expectations of science and its long- and short-term deliverables (Harwood, 2000, pp. 143-168). This is not to say that government parties were exclusively interested in 'applied' kinds of science. In a nationalist context, the cultural prestige of world-leading research served as an incentive for investment in 'pure' science, too. Indeed, some of the greatest German scientists of the time – the future Nobel laureates Robert Koch (1843-1910) and Paul Ehrlich (1854-1915), among others – worked in institutions co-sponsored by government and business parties, with rich results for 'knowledge' and 'profit' alike (Lenoir, 1988).

Assuming that Voller was not blind to such productive forms of cooperation, we may interpret his question about the 'aims of science' in Wilhelmine Germany more specifically as a question about the *effects* of business involvement or, in his case, government interference in science. How do the social, political, economic, and/or patriotic aims behind government funding of science interfere with the epistemic aims that Voller believed to be defining of science? How does this interference work out, in terms of the kind of research it promotes and, especially, the kind of demands it makes on scientists? What are the virtues or, perhaps, the vices that it encourages by pursuing other than epistemic aims?

Conclusion

When we fast-forward more than a century and examine how academics currently reflect on government interference in science, it is obvious that the language of virtue and vice that Voller's generation preferred has almost entirely disappeared – even though Steven Shapin observes that dispositions and character traits of the sort that used to be called virtues have become more important than ever in the 'new opacity' of the modern scientific-industrial complex (Shapin, 2008). Also, the language of 'purely scientific goals' has become suspect, as it draws too obviously on ideals of purity that we have been taught to distrust.[5] We have become accustomed to seeing science, past and present, 'as if it was produced by people with bodies, situated in time, space, culture, and society, and struggling for credibility and authority' – and hence as 'never pure' (Shapin, 2010). Consequently, it has become natural to assume that aims of science come in the plural and that this pluralism is inevitable, given the variety of interests pursued by the various profit and non-profit parties that are engaged in what is sometimes called 'Science 3.0' (Miedema, 2012).

Does this imply that Voller's question has become irrelevant, because his value-laden distinction between different aims of science fails to recognize that science as we actually know it is pursuing several aims at once? On the contrary: there is a sense in which Voller's question has become more relevant than ever. For if a pluralism of aims is the norm, then a well-developed sense of discernment is needed for seeing how these aims interact. Powers of distinction, indeed, are needed for assessing the relative weight of these aims. Are there circumstances under which some aims gain importance at the cost of others? Are there circumstances under which, say, the 'economic gaze' becomes so dominant that moral, aesthetic, political, or epistemic aims become subordinated? And if so, how desirable is that?

More specifically, Voller's question stimulates reflection on what sort of conduct is rewarded and thus encouraged by various 'constellations of goods' (combinations of aims that scientists pursue). What sort of scientific attitudes, ethos, or habitus are implied in, say, competitive research funding schemes or concentration of funding in 'top sectors' and 'research agendas'? How do changes in the constellations of aims that we pursue contribute to changing perceptions of what defines a 'successful' scientist? Or the other way around: if we spend increasing amounts of time monitoring our research output and its impact in terms of citations and Altmetric scores, which aims of science do we thereby serve? What do our practices and preoccupations reveal about the relative importance we attach to epistemic

---

5 On ideals of purity in late nineteenth-century Europe, see Arnold Labrie, *Zuiverheid en decadentie: over de grenzen van de burgerlijke cultuur in West-Europa, 1870-1914* (Amsterdam: Bert Bakker, 2001) and *De hang naar zuiverheid: de cultuur van het moderne Europa*, ed. by Rob van der Laarse, Arnold Labrie, and Willem Melching (Amsterdam: Het Spinhuis, 1998).

and non-epistemic aims of science? And what sorts of scientific personae, or ideal-typical models of what it takes to be a scientist, do we cultivate by the constellations of aims that we pursue?[6]

Questions about the aims of science are thereby inherently moral ones: they address core issues in what is called 'research ethics' or 'ethics of science'. This is not to say that they are driven by a specific moral agenda or premised on *a priori* suspicion of non-epistemic aims – even though it requires but little reflection to see that questions about the aims of science are most frequently raised out of concern and prompted by a perceived shift of emphasis from epistemic (knowledge) to non-epistemic aims (profit, reputation, influence). Indeed, our early 21st-century context resembles Voller's in that questions about the goods that scientists pursue are typically raised by critical commentators such as David B. Resnik, the author of *The Price of Truth: How Money Affects the Norms of Science* (2007), and authors belonging to the so-called 'slow science' movement, intent on challenging the capitalist regime of time as it manifests itself in academia (Resnik, 2007; Mountz et al., 2015, pp. 1235-1259). These voices, however, do not remotely have anything like a shared moral agenda, let alone a shared view on the relative weight that different aims of science deserve. What they share is a question and a general sense of dissatisfaction more than answers or solutions. As such, they invite us to join a moral quest more than to accept a pre-established point of view.

The question posed in the title of this chapter must therefore be read as, in principle, an open question: What are the goods or aims furthered by government interference in science? Also, it must be read as a context-specific question, in the sense that it cannot be answered in the abstract. Given that government interference can take different forms and be driven by different aims, deliberation has to focus on the specific shape that interference takes in specific historical circumstances and the effects it exerts on actual scientific practice. Just as Voller tried to weigh and judge the shifting constellations of aims that scientists in Wilhelmine Germany pursued, so we are invited to examine in our time and place how our aims and corresponding practices are being affected by new politics of interference and how desirable that is from a research ethical point of view.

---

6   This is the backbone of the ethics of historical scholarship I have been developing so far in Herman Paul, 'Weak Historicism: On Hierarchies of Intellectual Goods and Virtues', *Journal of the Philosophy of History*, 6 (2012), 369-388; 'What Could It Mean for Historians to Maintain a Dialogue With the Past?', *Journal of the Philosophy of History*, 8 (2014), pp. 445-463; and 'What Is a Scholarly Persona? Ten Theses on Virtues, Skills, and Desires', *History and Theory*, 53 (2014), 348-371.

References

Bud, Robert, '"Applied Science": A Phrase in Search of a Meaning', *Isis,* 103, 2012, pp. 537-545

Echterhölter, Anna, *Schattengefechte: genealogische Praktiken in Nachrufen auf Naturwissenschaftler (1710-1860)* (Göttingen: Wallstein Verlag, 2012)

Foot, Philippa, *Virtues and Vices and Other Essays in Moral Philosophy* (Berkeley, CA: University of California Press, 1978)

Glasser, Otto, *Wilhelm Conrad Röntgen und die Geschichte der Röntgenstrahlen*, 2nd ed. (Berlin; Göttingen; Heidelberg: Springer, 1959)

Gooday, Graeme, '"Vague and Artificial": The Historically Elusive Distinction between Pure and Applied Science', *Isis,* 103, 2012, pp. 546-554

Harwood, Jonathan, 'Das Selbstverständnis des Naturwissenschaftlers im Wandel: die Lage innerhalb und außerhalb der Akademie zu Beginn des 20. Jahrhunderts', in *Die Preußische Akademie der Wissenschaften zu Berlin, 1919-1945,* edited by Wolfram Fischer (Berlin: Akademie Verlag, 2000), pp. 143-168

Kevles, Daniel J., 'The Physics, Mathematics, and Chemistry Communities: A Comparative Analysis', in *The Organization of Knowledge in Modern America,* edited by Alexandra Oleson and John Voss (Baltimore, MD: Johns Hopkins University Press, 1979), pp. 139-172

Kline, Ronald, 'Construing "Technology" as "Applied Science": Public Rhetoric of Scientists and Engineers in the United States, 1880-1945', *Isis,* 86, 1995, pp. 194-221

Labrie, Arnold, *De hang naar zuiverheid: de cultuur van het moderne Europa,* edited by Rob van der Laarse, Arnold Labrie, and Willem Melching (Amsterdam: Het Spinhuis, 1998)

Labrie, Arnold, *Zuiverheid en decadentie: over de grenzen van de burgerlijke cultuur in West-Europa, 1870-1914* (Amsterdam: Bert Bakker, 2001)

Laudan, Larry, *Science and Values: The Aims of Science and Their Role in Scientific Debate* (Berkeley/Los Angeles/London: University of California Press, 1984)

Layton, Jr., Edwin T., 'American Ideologies of Science and Engineering', *Technology and Culture,* 17, 1976, pp. 688-701

Lenoir, Timothy, 'A Magic Bullet: Research for Profit and the Growth of Knowledge in Germany Around 1900', *Minerva* 26, 1988, pp. 66-88

Lucier, Paul, 'The Origins of Pure and Applied Science in Gilded Age America', *Isis,* 103, 2012, pp. 527-536

Lucier, Paul, *Scientists and Swindlers: Consulting on Coal and Oil in America, 1820-1890* (Baltimore, MD: Johns Hopkins University Press, 2008)

Miedema, Frank, *Science 3.0: Real Science, Real Knowledge* (Amsterdam: Amsterdam University Press, 2012)

Mountz, Alison, et al., 'For Slow Scholarship: A Feminist Politics of Resistance through Collective Action in the Neoliberal University', *ACME* 14, 2015, pp. 1235-1259

Neubert, Franz (ed.), *Deutsches Zeitgenossenlexikon: biographisches Handbuch deutscher Männer und Frauen der Gegenwart* (Leipzig: Schulze & Co., 1905)

Paul, Herman, 'Weak Historicism: On Hierarchies of Intellectual Goods and Virtues', *Journal of the Philosophy of History*, 6, 2012, pp. 369-388

Paul, Herman, 'What Could It Mean for Historians to Maintain a Dialogue With the Past?', *Journal of the Philosophy of History*, 8, 2014, pp. 445-463

Paul, Herman, 'What Is a Scholarly Persona? Ten Theses on Virtues, Skills, and Desires', *History and Theory*, 53, 2014, pp. 348-371

Paul, Herman, *Key Issues in Historical Theory* (New York; London: Routledge, 2015)

Paul, Herman, 'Virtue Language in Nineteenth-Century Orientalism: A Case Study in Historical Epistemology', *Modern Intellectual History* (forthcoming)

Remsen, I., 'American Influence in Civilization', *The Popular Science Monthly*, 13, 1878, pp. 495-497

Remsen, I., 'Thoroughness', *Williams Review*, 3, 1872, pp. 33-34

Resnik, David B., *The Price of Truth: How Money Affects the Norms of Science* (Oxford: Oxford University Press, 2007)

Rowland, H.A., 'A Plea for Pure Science', *Proceedings of the American Association for the Advancement of Science*, 32, 1884, pp. 105-126

Saarloos, Lejón, 'Virtue and Vice in Academic Memory: Lord Acton and Charles Oman', *History of Humanities* (forthcoming)

Shapin, Steven, *The Scientific Life: A Moral History of a Late Modern Vocation* (Chicago/ London: University of Chicago Press, 2008)

Shapin, Steven, *Never Pure: Historical Studies of Science as if It Was Produced by People with Bodies, Situated in Time, Space, Culture, and Society, and Struggling for Credibility and Authority* (Baltimore, MD: Johns Hopkins University Press, 2010)

Thurston, Robert H., 'The Mission of Science', *Proceedings of the American Association for the Advancement of Science*, 33, 1885, pp. 227-253

Toulmin, Stephen, *Foresight and Understanding: An Enquiry into the Aims of Science* (London: Hutchinson, 1961)

Voller, August, *Ueber Aenderungen der electromotorischen Kraft galvanischer Combinationen durch die Wärme* (Göttingen: Dieteriseh'schen Univ.-Buchdruckerei, 1873)Voller, August, 'Physikalisches Kabinett', *Jahrbuch der Hamburgischen wissenschaftlichen Anstalten*, 1 (1884), lxvii-lxxi

Voller, A., 'Physikalisches Staats-Laboratorium', *Jahrbuch der Hamburgischen wissenschaftlichen Anstalten*, 3 (1886), p. lxiii-lxv

Voller, A., 'Physikalisches Staats-Laboratorium', *Jahrbuch der Hamburgischen wissenschaftlichen Anstalten*, 9 (1891a), pp. lxxi-lxxiv

Voller, A., 'Zur Erinnerung an Wilhelm Weber', *Zeitschrift für den Physikalischen und Chemischen Unterricht,* 5 (1891b), 29-30

Voller, A., 'Physikalisches Staats-Laboratorium', *Jahrbuch der Hamburgischen wissenschaftlichen Anstalten,* 12 (1894), pp. lxii-lxiv

Voller, [A.], 'Das Physikalische Staats-Laboratorium', in *Hamburg in naturwissenschaftlicher und medizinischer Beziehung: den Teilnehmern der 73. Versammlung deutscher Naturforscher und Ärzte als Festgabe gewidmet,* ed. by [J.] Classen and [Th.] Deneke (Hamburg: Leopold Voss, 1901), pp. 205-212

Weinmeister, P. (ed.), *J.G. Poggendorffs biographisch-literarisches Handwörterbuch für Mathematik, Astronomie, Physik, Chemie und verwandte Wissenschaftsgebiete,* vol. 5 (Leipzig; Berlin: Chemie Verlag, 1925-1926)

Witte, Karl, 'Zur Geschichte des Physikalisches Staatsinstituts und der Physik in Hamburg', in *100 Jahre Physik in Hamburg,* ed. by Klaus Tornier (Hamburg: Universität Hamburg, 1985), pp. 9-27

Author biography

Herman Paul is associate professor of historical theory and historiography at Leiden University, where he directs a research project on 'The Scholarly Self: Character, Habit, and Virtue in the Humanities, 1860-1930'. He also holds a special chair in secularization studies at the University of Groningen and is an elected member of The Young Academy (KNAW). Recent book publications include *Key Issues in Historical Theory* (Routledge, 2015) and *Hayden White: The Historical Imagination* (Polity, 2011).

Free-range Poultry Holdings

Living the Academic Life in a Context of Normative Uncertainty

Beatrice de Graaf[1]

These days, we see increasing numbers of scholars aspiring to live the scientific life: longing to join academia, hoping to follow their vocation, to make a career here and hone their theoretical skills to perfection. At the same time, uncertainty regarding life as an academic is mounting. This uncertainty may be enforced by the fact that these young scholars are drawn into an unwanted process of (self-)selection. Although the majority of these young scholars would like to remain in academia, the fact is that for every ten there is room in the university for only one or two of them. Research potential surpassing the available budget – this dynamic tends to reduce autonomy, liberty of choice and diversity within the research environment. And young scholars are amongst the first to be exposed to this worrisome trend.

In this chapter, I will present two narratives that seek to outline the academic life and its purpose: the utilitarian 'goose model' and the Humboldtian '*Bildung* model'. We will see that the ideas, goals, and expectations of each model continue to compete for recognition and endorsement. And although one of the two is undoubtedly gaining the upper hand, the values of the other model are essential to sustaining the life of the mind. This conflict of values regarding science and the scientist is precipitating a significant degree of uncertainty in politics, academia, and society regarding the aspirations of the academic endeavour and the norms that (should) hold for these. Students, scholars, and administrators are uncertain about how to act given the diversity of moral doctrines, about how to decide which moral conviction applies when and how – and based on which criteria and whose authority. Our theoretical pursuits are at stake, but who is entitled to decide how best to protect and promote them?

---

1   *Many thanks to Christoph Baumgärtner, Maarten Prak, and Ingrid Robeyns for their comments and suggestions. This text is the adaptation of the lecture at the opening of the Academic Year at the Erasmus University Rotterdam in September 2015.*

The golden goose

The first story's opening is very familiar: 'Once upon a time there was a goose who laid a golden egg every day'.² In 2015, Director-General for Research and Innovation of the European Commission, Robert-Jan Smits, passionately argued to keep the EU's research investment programmes afloat also during times of financial crises. In his words, it would be very unwise to 'kill the goose that will lay golden eggs in the future'. He underscored his admonition by pointing to Finland, which overcame its economic crisis in the 1990s by increasing investments in innovation and research, and to Germany, which has been hitting the ceiling with an extra 18 billion euros for research since the financial crisis began. Compared to these efforts, the EU as a whole does not strike an impressive figure: notwithstanding the common European goal of investing 3 percent of GDP in research, today's figure currently amounts to a meagre 1.9 percent. 'We cannot build a knowledge society if we don't invest in it', says Smits (EU, 2015).

What is interesting here is the language in which Smits' plea is couched and the urgency with which it was made. Using a metaphor taken from Aesop's fable about the goose who laid golden eggs, Smits was urging policymakers, investors, and bankers, even the EU as a whole, to see the current situation in perspective. The scientist as goose, or as the egg, is a powerful narrative, easily grasped, and most probably designed to reach and win the hearts and minds of the power wielders in Europe. Even they should be lured, captivated, and plied by the shine of the golden eggs, and thus refrain from slaughtering or starving the goose. In its crudest form, science and scientists are here to make money, to increase GDP (*Het Financieele Dagblad*, 2015).³ Or, in a more benevolent version of the same tale, they

---

2   "A man and his wife owned a very special goose. Every day the goose would lay a golden egg, which made the couple very rich.
'Just think,' said the man's wife, 'If we could have all the golden eggs that are inside the goose, we could be richer much faster.'
'You're right,' said her husband, 'We wouldn't have to wait for the goose to lay her egg every day.' So, the couple killed the goose and cut her open, only to find that she was just like every other goose. She had no golden eggs inside of her at all, and they had no more golden eggs." http://www.storyit.com/Classics/Stories/goldengooseegg.htm

3   Around mid-August 2015, 'Brussels' subsequently announced that it was going to develop novel macro-economic models to better monitor and evaluate the net return of its R&D investments – current economic models consider research and innovation as debit items, with returns projected too far into the future to be calculable, and therefore to be excluded from the credit side. 'EU laat impact innovatie op economie onderzoeken'. *Het Financieele Dagblad*, 11 August 2015.

are expected to solve the problems of humankind: to produce more and healthier food, cure cancer, fight climate change, increase sustainability, and help to achieve the millennium development goals. Indeed, these are all essential values. Take for example Jan Tinbergen, the first and thus far last Dutch winner of the Nobel Prize for Economics. He explicitly subscribed to this utilitarian 'goose model'. For him, government spending on science and education was essential, since they directly contributed to reductions in income inequality. Science policy ought to be designed to reduce income inequality – a veritable blast from the past (Van Rompuy, 1974, p.66).

This functional, or common-sense, 'goose model' underlies many of our academic and research agendas, as well as many NWO programmes, and it most certainly informs the so-called spearhead programme, or top sector policy ('topsectorenbeleid'). There are, however, more stories to tell than this particular Greek fable. From the following narrative, originating in Germany, the uncertainties and clashes about goals and norms that emerge for governments and universities from these stories' incongruences will be explained.

Bildung

This powerful story has recently been enacted in the streets of Amsterdam, where students and staff have gained a certain amount of notoriety protesting against the withering away of the *communitas academica.* Demonstrators, there and elsewhere, were objecting to the strict production and output standards that have been inflicted upon them (and us) over the last decade, when already insufficient budgets were being usurped for campus real estate projects right under their noses. Some of the idealistic rebels were inspired by a longing for the classic ideal of the university as a sanctuary for passionate professors, intellectual interlocutors, and freethinking spirits; for the university as a site for 'disinterestedness' (Robert K. Merton) (Macfarlane and Cheng, 2008).

These protests have been a powerful reminder of the second narrative that can be told about the world of higher education. We could call it the story of *Bildung*; not so much a fairy tale by the Grimm brothers but rather a path with Humboldtian roots. In this story, the university is a place where norms, values, ethics, and ideals are developed, cultivated, and discussed between students and teachers. In the words of our very own Minister of Education: 'Universities and institutions of higher learning are training our future leaders. Teachers, judges, nurses, and architects alike – people who set the tone for how we engage and deal with each other in our society' (*De Volkskrant*, 2015).

Martha Nussbaum's *Not for Profit: Why democracy needs the humanities* develops this story further (2010). Her plea for *Bildung* offers a model that does not provide us with one-way research paths culminating in clear-cut outcomes – in this case, the 'eggs'. It is a model that particularly values the education of critical and empathic citizens and seeks to equip scholars with critical tools to set out on different routes and in different directions. *Bildung* does not tie in so well with the logic of the neoliberals or the grammar of a capitalist economy; it is rather a model of critical pedagogy for developing individual responsibility, pioneering innovation, and the self-examination of democratic citizens. This model presupposes an open and liberal society; one that does not tell researchers what to do, or at least does not dictate the diversity or direction of their inquiries in detail. Instead, this model challenges and enables academics to use their specific capacities for contributing to the common good, by, for example, monitoring the ethical priorities, normative proclivities, and professional skills of researchers in terms of scientific integrity, or by assessing their ability to teach '21st-century skills'. It considers universities as 'archives of our common knowledge', critical caretakers of the public good, as, in the words of Ingrid Robeyns (in her inaugural address), 'centres for independent thought' (Robeyns, 2015; Hutchins, 2015, pp. 53-54), and as a community or *civitas* where new citizens, ideas, inventions, and potentially innovative initiatives circulate (Schinkel, 2015, pp. 53-54).

Normative uncertainty

Having briefly described these two models, it is necessary to emphasize that *both* are valuable (perhaps even equally valuable). As a historian, I was trained in the ideals of the humanities as expressed by Humboldt and Nussbaum, both at Utrecht and Bonn University, and drilled in the German way of questioning and deconstructing definitions and concepts such as security, terrorism, and democracy. Students in political science or history are still trained to study what sets a democratic charter apart from totalitarian repression – and how easily lapsing into state terror can happen; exactly the kind of insights that Nussbaum wants scholars in the humanities to develop. On the other hand, if I may draw from my own experience as a researcher, while working at Leiden University's Centre for Terrorism and Counterterrorism I experienced true satisfaction from building concrete terrorism databases and evaluating counterterrorism laws – directly contributing to a common good (in this case security), rather than devoting myself to indirect, self-reflexive critique alone.

The uncertainty mentioned earlier comes into play when we are confronted with a plurality of values and purposes underlying our diverse ideas about scientific life – *and when we have to make choices and don't know how to handle this incongruity.* Do we need to develop better antiterrorism equipment, or should we concentrate more on *understanding and critiquing* the advent of the surveillance state? Many researchers in the humanities and social sciences experience this ambivalence first-hand; that they are torn between these two ways of thinking, seeing them both as valuable. Few of them are probably willing to commit wholeheartedly to only one of these. Few of them would want to retreat completely into the ivory tower. Most of them are willing to make a contribution – directly or indirectly – to the improvement of society, but feel uncomfortable when their work is being completely reduced to this contribution alone.

This conflict of values and purposes, and the uncertainty that often ensues, has gained more salience in the current situation of budgetary constraints caused by the current state of economic and financial crises and cutbacks in government spending. So-called top sectors, spearheads in innovation and research, have been designated and research monies rechannelled into industrial budgets (Valkema, 2015). The NWO, which is one of the main pillars of the (highly productive) Dutch scientific biotope, is going through a process of restructuring. Researchers have increasingly come to rely and depend on large-scale EU programmes, but success rates are declining dramatically, from 25 to 10-16 percent or even lower (Floratos, 2015). In 2015, the NWO success rate in the humanities even touched a disappointingly dismal low of 7.5 percent. Although the NWO is meant to support the natural development of science itself and to accommodate the rise of multi-and interdisciplinary approaches, these dwindling success rates leave the impression within the academic world, especially amongst young researchers, that they hardly stand a chance to launch a career in research. On top of that, the massification, commodification, specialization, and internationalization of higher education (Nowottny et al., 2002; Kerr, 2001; Stolker, 2014) have all left their mark on Dutch universities as well.

Against this backdrop of scarcity, the state of uncertainty occasioned by the plurality of values and purposes is highlighted even more and is too often transformed into relentless competition, which starts to spark real conflicts. In other words, one of these narratives, the common-sensical goose model, has started to 'colonialize' the social subsystem of science by beginning to evaluate it with the logics of a different social subsystem, that of the economy. Utilitarianism in its crude economic form is becoming the dominant discourse, in society as well as in academia. Scholars and universities are being pushed to the assembly line, pressed to produce preconceived eggs. And it is exactly this imbalance that is troubling. Since the 1960s, Dutch universities' budgets increased, and academia flourished as it provided room for cooperation and competition between scholars, research schools, and universities. Nowadays, the dynamics of competition often prevent any long-term investment in cooperation and are undermining the egalitarian model of this productive Dutch scientific ecology (Prak, 2009).[4] Of course, academics understand the futility of a state of absolute non-interference from the outside. They often benefit significantly from external support to finance, develop, and apply their research. They want to be in touch with society, partly because urgent social problems prompt new research questions. Large-scale infrastructures and laboratories, PhD training and hiring schemes often need to be developed 'from above'; programmes in minor languages require protection and funding (it is a pity none of them submitted research questions to the Dutch National Research Agenda!). But it is an illusion to believe that someone from outside or above can design in advance the next (sort of) 'golden egg' or the facility needed to produce it. Even if such a programme of academic engineering would be successful in achieving particular goals, it would not be conducive to new and surprising developments and outcomes.

In short, the problem is not the plurality of values and purposes itself, but the attempt by one of these models to overwhelm all of the other visions of academic life that our open, liberal, and pluralistic society has to offer.

The academic life

The first step to restore this imbalance is to acknowledge and defend the diversity and richness of the academic lives at stake here, and to counter moves that might have one vision monopolize all others. Many dedicated academics, university boards, and organisations, like The Young Academy, have already made this agenda a priority.

Academic life cannot be regulated from above. Scholars do not stand orderly in line – not in real life, and not in history. Science is never tidy, unified, or simple. Academics live in a multiform community. Some academics adopt

---

4   Interview with Hans Clevers, *Maarten!*, April/May 2015, p. 47.

the role of modern-day prophet, moral commentator, or priest in public service, unleashing warnings about levels of pollution, climate change, and terrorism, or drawing attention to social fissures, sometimes even courted as charismatic truth speakers in a world of uncertainties (Shapin, 2008, xv). Other scientists work hard to plan, secure, engineer, develop, and cultivate natural and social environments. They comment on migrant streams, research brain development, or improve economic education. Still others serve science and society alike in their laboratory, for example to map and identify new viruses.

All these scholars belong together, in one university and one academic community. Selection and prioritization does not enrich the flock's environment, it only impoverishes it. Different scholarly personae are, according to Herman Paul, 'characterized by different constellations of virtues and skills or, more precisely, by different constellations of commitments to goods (epistemic, moral, political, and so forth), the pursuit of which requires the exercise of certain virtues and skills' (Paul, 2014, p. 348). Instead of prescribing outcomes, results, and products, what would really be beneficial is aiding and abetting these skills. Any story about academic life has to commit itself to watch over this invaluable academic ecosystem (Knottnerus) and to shield it against any attempt to tear it apart.

Tend the flock

Our interlocutors past and present – Aesop, Humboldt, Nussbaum, and others – have enough advice to offer to help us come to terms with the normative uncertainty that renders our lives so complicated today. Based on their stories, a case could be made to improve the balance between the two models outlined above – not to defend well-vested interests or privileges, but to protect the reality that academic space and variety are 'necessarily instrumental' (Robeyns, 2015) to keep academia alive and have it serve society as it should. Here are some suggestions to help create more space for diversity in academic life, and to facilitate a 'balance of power' by protecting the *Bildung* model from questionable preferences for the goose model:

- Protect the young geese. A sustainable research environment is all about stimulating young talent and enabling untied research (in the Netherlands: increase the budget of NWO's *Vernieuwingsimpuls*).
- Tend the flock. Knowledge bearers do not dwell well alone. Rat-race dynamics increases stress and wears down flock fertility; whereas a Brady Bunch of scholars of all kinds of feathers and colours will aid

fecundity nicely. Ergo: Increase the number of individual PhD positions at the department level, rather than embed them in large-scale grant programmes at the national or international level.
- Feed the flock. Ergo: Sow seed money with broad strokes to stimulate diversity and surprise, with no advance prioritization or selection of disciplines or themes.
- Feed generously. EU's standards of scientific funding have fallen below the 3 percent mark in both the Netherlands and other countries. Grant success rates have to surpass the 18 percent mark in order to prevent research from becoming a lottery.
- Organise free-range poultry holdings. In line with Ingrid Robeyn's investigations into the workload of academics (Robeyns, 2015)[5], give researchers the time they need to think and to write, and to take time out now and then. Ergo: Implement the Anglo-American sabbatical whereby every six semesters with a regular teaching load, upon approved application, one is entitled to one sabbatical semester of research.
- Don't discriminate. Universities need talented teachers as much as they do research geniuses, media darlings, and administrators. Ergo: Encourage public and academic service by conferring awards (yes, with a monetary incentive) for good teaching, scholarly achievement, and media presence.
- Let the geese loose. Ergo: Stick to the Haldane Principle, i.e. accommodate the idea that decisions about what to spend research funds on should be made by peers rather than by managers or politicians (cf. Kan, 2014).[6]
- Acquaint others with the flock: Sell first row seats to politicians, managers, and captains of industry, allowing them to contribute to lectures or to enjoy a research internship within research groups or laboratories – in order to demonstrate the value of the *Bildung* model from within.

Group portrait with scientists

Hopefully, the National Research Agenda (*Nationale Wetenschapsagenda*, NWA) will be able to highlight and help to protect the varieties of and diversities within academic life in the Netherlands. The Ministry of

---

5   This is her – highly timely – appeal to social scientists to launch statistical investigations into scientists' workloads in the Netherlands.
6   This principle is named after Richard Burdon Haldane, a British official who in 1904 and from 1909 to 1918 chaired committees and commissions that recommended this policy.

Education launched this plan to facilitate links and connections between various research agendas and to help identify pressing questions posed by the Dutch populace that deserve further research. The chairs of the NWA Steering Committee think that it is also important to turn this initiative into a platform that demonstrates what scholars here in the Netherlands are already capable of – and why they need and deserve more resources. The NWA has been able to showcase the wealth of questions coming from the general public, as well as to suggest possible ways in which Dutch scholars can best address these questions. The NWA calls for diversity within and protection of our academic ecosystem, not just to produce more 'eggs', but also – in line with Nussbaum – to enhance our society's ability to think critically, to educate knowledgeable and empathetic citizens and to deal with complex global problems.

Moreover, and perhaps most importantly, the NWA functions as a vehicle for combining the two above outlined narratives, it serves as a platform for connecting different types and approaches of and to research. Among the 12,000-plus queries submitted online, many asked questions having to do with the origins of mankind, with society's resilience, with identity, democratic citizenship, and the need for spirituality and religiosity. Utilitarian motives did not predominate. The NWA explicitly intends to honour these pressing questions, and where possible will help appropriate parties to rise to this challenge.

Germany we are not – spending 18 billion euros and buying off critique from the humanities and other corners. But perhaps subsequent government coalitions could take a look at Finland, a small country that managed to seriously invest in a knowledge society even in times of severe crisis. Golden eggs rolling off an assembly line might speak to some, but coming to knowledge does not, nor do those who harbour, garner, and cultivate its growing belong to the realm of fairy tales. Neither should these fertile 'geese' be confined to/by large-scale poultry halls.

In this chapter, a case has been made for regulations protecting and enabling knowledgeable, informed, well-staffed, and knowledge-seeking communities – laboratories, research and development departments, and universities alike (as heterogeneous, complex, and multiple as they may be). Historian Lorraine Daston argued that the history of science provides a unique self-portrait of Europe. She said that 'no other culture has relied so heavily on the history of science to define its own identity. Since Europe became Europe in its own eyes, science has been held up as its image and its emblem – whether understood as inexorable progress of vertiginous change or tragic loss of tradition' (Daston, 2005, p. 30). Society would do well to harbour and nourish variety within academic life, and uphold an openness and correlative degree of unpredictability, regarding the plurality of goals and purposes that need to be retained within the halls of the academy.

References

'Bussemaker heeft futuristische visie', *De Volkskrant*, 8 July 2015

'EU laat impact innovatie op economie onderzoeken', *Het Financieele Dagblad*, 11 August 2015

Daston, Lorraine, *The History of Science as European Self-Portraiture*, Preamium Erasmianum Essay 2005 (Amsterdam, 2005)

EU 2015, 'Don't talk! Invest!, says new Director-General of the EU's Research DG', http://cordis.europa.eu/news/rcn/122987_en.html, 28 May 2015 (retrieved 11 August 2015)

Hutchins, Robert, 'The Freedom of the University', *Ethics*, 61(2), 1951, pp. 95-104

Kan, Alexander Rinnooy, 'Met de kennis van straks', *KNAW lecture*, 26 May 2014

Kerr, Clark, *The Uses of the University* (Cambridge, MS: Harvard University Press, 2001)

Macfarlane, Bruce, and Ming Cheng, 'Communism, Universalism and Disinterestedness: Re-examining contemporary support among academics for Merton's scientific norms', *Journal of Academic Ethics*, 6, 2008, pp. 67-78

Nowottny, Helga, Peter Scott, and Michal Gibbons, *Re-Thinking Science: Knowledge and the public in an age of uncertainty* (Cambridge, MS: Harvard University Press, 2002)

Nussbaum, Martha C., *Not for Profit: Why democracy needs the humanities* (Princeton: Princeton University Press, 2010)

Paul, Herman, 'What Is a Scholarly Persona? Ten Theses on Virtues, Skills, and Desires', *History and Theory* 53, 2014, pp. 348-371

Prak, Maarten, 'Bericht uit de halvarinefabriek. Nederland kennisland? (2)', *De Groene Amsterdammer*, 3 June 2009

Robeyns, Ingrid, 'Universiteit is voor kritisch en onafhankelijk denken', *DUB*, 10 April 2015

Robeyns, Ingrid, 'Waarom een lage werkdruk zo belangrijk is', *DUB*, 16 March 2015

Rompuy, E. van, *Jan Tinbergen. De eerste Nobelprijswinnaar economie* (Antwerpen/Utrecht: Het Spectrum, 1974) p. 66.

Schinkel, Willem, 'Wat zijn de publieke taken van de universiteit?', *Beleid en Maatschappij* 42 (1), 2015, pp. 51-54

Shapin, Steven, *The Scientific Life: A moral history of a late modern vocation* (Chicago/London: University of Chicago Press, 2008)

Stolker, Carel, *Rethinking the Law School: Education, research, outreach and governance* (Cambridge 2014)

Valkema, Fridus, 'De schade van de topsectoren', *Technisch Weekblad,* 14 August 2015

A National Research Agenda and the Self-understanding of Modern Universities

Rutger Claassen and Marcus Düwell

Introduction

The process of establishing a National Research Agenda (*Nationale Wetenschapsagenda*, NWA) that has been undertaken in the Netherlands in 2015-2016 seems to have been both a continuation and a break with the recent past in science policy. A continuation insofar as it fits with the trend of channelling research funds increasingly to certain *communally* prioritized research topics, instead of leaving this act of priority-setting to *individual* researchers. Unsurprisingly, this has elicited the usual objections by those who cherish the ideal of 'academic freedom' for the individual researcher. However, it seems to have been a break with this trend in that the community deciding the priorities was not the community of scientists themselves, nor organised private interests or private-public partnerships (as has been the case with the establishment of the *Topsectoren*, or NWO programmes like *Socially Responsible Innovation*). Instead, a radically bottom-up process was organised in which the public at large could pose its questions to scientists. One might have had the impression that this would be a moment of radical innovation, in which 'democracy meets science'.

However, appearances can be deceiving. In this chapter, we will first argue that the NWA is in fact primarily a continuation of the older practice of giving organised private interests a firmer grip on research priorities. The 'democratization' of science policy is just one of four possible justifications for setting up an NWA which we will distinguish in this essay (section 1). This raises the question of which of these justifications are justified. How should academic research be organised? And what is the role of a national research agenda in academic research?

To answer these questions, we will first give a philosophical analysis of the role of the university. Scientific research at its core is about the self-understanding of human beings. Academic researchers help individuals and groups to deal with the fundamental challenges of human existence and uncover possible perspectives on what it is to lead a (good) human life (section 2). From this perspective, we will argue that academic research needs both *reflective distance* from social actors and processes and *reflective connectivity*. Researchers need to guard their independence from society to be creative and reliable, and to develop long-term perspectives. At the same time, researchers need to establish meaningful connections to social actors and practices, because it is their human challenges and self-understanding which are ultimately at stake in research (section 3). Given these fundamental reflections, we formulate seven recommendations for agenda-setting in research policy. A thread which runs through all of these is a concern with the direct steering role that is currently given to a select group of private actors. Our plea is for a stronger steering role for the academic community itself, while at the same time establishing stronger, more permanent, and more in-depth discussion fora with a wider group of societal stakeholders, both public and private, commercial and non-profit (section 4).

Setting up the National Research Agenda in the Netherlands

The NWA experiment for Dutch universities has received much publicity. After a high-level public announcement (with the chairpersons of the NWA appearing in TV show *De Wereld Draait Door* and other media outlets), it began with a first round in which a significant number of possible research questions, close to 12,000, were sent in. In a next stage, these were ordered into 140 aggregated questions. On the basis of these questions, various clusters of questions – so-called routes – were formed. Examples of these are 'Resilient and Meaningful Societies', 'Brain, Cognition and Behaviour', 'Circular Economy' and 'The Sustainable Production of Healthy and Safe Food'. In the organisation of these latter stages, the so-called 'knowledge coalition', formed by representatives of various research-related parties – among them the Netherlands Organisation for Scientific Research (NWO) and the Royal Academy of Arts and Sciences (KNAW), but also representatives of industry (VNO-NCW) and technology (TNO), played a leading role. In the follow-up to this process, bigger research activities have been planned under the direction of this knowledge coalition. These research activities bring together financing from the NWO and various societal and economic partners in order to form large, aggregated research activities. The effect of the clustering by the knowledge coalition will likely be that a significant amount of research money will be invested in specific directions.

The entire experiment was accompanied by a number of formal and informal meetings and discussions. Faculties, departments, research schools, and other academic bodies discussed how to relate to this agenda. These discussions were a continuation of debates at Dutch universities over the last few years, as a consequence of recent trends. *First*, there was the decision of the Dutch government of 2008 to shift funds from structural financing of the universities to heavy competitions organised at the NWO. *Second*, the knowledge coalition formulated in 2011 a policy that forced Dutch research organisations to invest a significant part of available research money in areas of industry that have had a specific relevance for Dutch economy in the past (*Topsectoren*). *Third,* there was a ministerial plan published in December 2014 to reorganise the NWO in such a way that barely any systematic form of self-governance in research would survive.

Discussions about research policy received new dynamics due to the protests in Amsterdam (the occupation of the Maagdenhuis) in early 2015 about the future of the Humanities and freedom in academic research. In this context, the NWA was perceived by many scholars as being yet another way of reducing universities to instruments for technological and economic innovation. Many researchers were critical and fearful of these developments. Two things were particularly striking: on the one hand, there seems to be a broad consensus amongst Dutch researchers about the importance and value of letting research follow its own logic. Only a university that can develop according to the internal dynamics of academic debate can flourish in terms of academic excellence, and can likewise respond appropriately to the needs of modern societies. On the other hand, however, there was a lack of convincing narratives about the legitimacy of such free research and about what appropriate decision procedures concerning the formulating of research priorities would be.

Against this background, the NWA plan was launched. Why would we want to have such an agenda? Of course one can have all kinds of suspicions as to which partners may be motivated to want such an agenda; one can speculate that some political parties want industry to have more influence on the research process, and one can speculate as to what the motivation of industry could be. We are aware that all kinds of interests will be at stake. But as philosophers we must discuss the legitimacy of such a process not on the basis of speculation concerning possible power interests, but rather we must first of all analyse critically whether or not there could be legitimate reasons for such a process. Various players within the process have made statements about the rationale or justification of having such an agenda. We will not reconstruct them here, but will summarise this by setting out four possible answers to the question, 'Why might we want to have an NWA?' (we cannot come up with more possible rationales for the legitimization of the research agenda, but we would be curious to learn whether there are more options).

A *first* possible aim in having a national research agenda could be a *heuristic* one. The public will come up with questions that researchers themselves have not thought of. It is quite probable that researchers and society at large have many interesting questions which are not addressed in current research; in fact, it would be surprising if this were not the case. So, in that sense, there could be reasons to find out what kind of questions the public may come up with. In relation to this, it may be valuable to make an *inventory* of what different stakeholders consider to be relevant research questions and to create links between them. We could easily see the research agenda as an attempt to form a forum in which the different research interests of various stakeholders and private persons are *articulated and related* to each other. One can, of course, wonder whether the enormous time pressure under which questions had to be articulated was helpful in facilitating the formation of well-formulated questions by stakeholders. In any case, we could see the NWA as a tool for mobilizing creative ideas about future research.

A *second* possible aim could be to *mobilize additional financing* for research. From this perspective the research agenda serves a merely strategic bargaining purpose. If the public come up with wildly attractive research ideas, government or private parties (such as big companies), so it is hoped, will make extra funds available. Only time will tell whether additional funding can be mobilized; at the moment it seems unlikely that government will be willing to invest more. Therefore, the hope now is primarily that there will be more substantial contributions from industry. However, this raises the problem of what the price of such co-financing from industry will be. If indeed such co-financing implies that industry will determine the research policy of the public funding of research, the price is very high. This problem leads directly to considering the next possible aim.

A *third* aim could be to organise a *more democratic process* of decision-making about research priorities. When the process began, some hoped (while others feared) that it would take a direct-democratic form. Indeed, everybody with an internet connection and basic Dutch language skills could submit questions. However, further in the process of forming the agenda, democracy does not play any role. The clustering of questions has been done by researchers and the follow-up activities are determined by the 'knowledge coalition'. The process is democratic insofar as the democratically legitimized Minister has initiated and approved the whole procedure, but no relevant democratic body has played a significant role in the further process. There was, for example, no relevant contribution from Parliament. Moreover, the knowledge coalition is not representative of civil society at large (i.e. representing different social or cultural organisations and interest groups). A broader plurality of possible stakeholders (such as representatives of artists, nature conservation organisations, advisory bodies etc.) were only involved in smaller advisory functions.

A *fourth* and final aim of the NWA could be to organise the process of priority setting in a more *transparent* way. Citizens were submitting questions, researchers were validating and clustering them, and under the guidance of the knowledge coalition but with the involvement of researchers and private partners, those clusters were transformed into research programmes. This transparency, however, is not based on in-depth analyses in terms of research desiderata, research excellence, or international research trends. Thus, one can wonder to what extent such a narrative would enhance the quality of decision-making concerning research priorities.

This concludes our brief overview of different expectations people had of the NWA. Have we missed a possible legitimating narrative? And what should we think of these expectations? Which of them are justified? Should we set research priorities democratically, or merely see such an agenda as having a heuristic or strategic goal? In an attempt to find answers to these questions, we must take a step back and first ask ourselves what the legitimate goals of research could be. This leads us to an inquiry into the 'philosophy of the university'. Only then can we come back to the more practical questions.

The philosophy of academic research: practical self-understanding

The question of what the relevance of academia is has preoccupied philosophers ever since something like methodologically organised forms of research have existed. Taking a short-cut, we can distil three valuable key aims of academic research.

First, the relevance of research lies in providing us with *solutions for the fundamental challenges of human existence*. Human beings have needs and, being dependent on their social and physical environments, may face many challenges in their lives. Research may help them to deal with those challenges: from the physical need for food and shelter, to combatting illnesses and resolving scarcities of energy, water, and other resources. Research in technology and medicine may seem to be of primary importance for this aim, but in fact the picture is much more diverse and complex. We do not know beforehand from which angle the most promising solutions can be expected to arrive. Problems of scarcity can be solved by technologies, but often solutions can also come from changes in human behaviour. Whether or not a specific technology is genuinely a solution for our problems will often depend on the cultural and political circumstances in which it is applied. Some fundamental technical changes (e.g. in the life sciences) need decades of research before they come to technical applicability. It is hard to predict to what extent those technologies really will provide significant solutions, or if they will raise more problems than they solve. Significant components of research today deal with solutions for the 'side-constraints' and consequences of earlier solutions. Prolonging life expectancy through better healthcare systems produces overpopulation. Increases in consumption, energy use, and emissions result in climate change. Whether or not the most significant solutions for all of these problems will come from the life sciences, from ICT, or from changes in the institutional and social setting we cannot say in advance. Perhaps the energy problem will be solved by the development of more sustainable forms of energy, perhaps by digital technologies that will make travelling superfluous, or perhaps we will simply change our habits in fundamental ways. It seems probable that solutions will be a combination of all these factors. However, since we do not know where solutions can be expected to come from, and since most of these research activities are premised on long-term investments, often on a global scale, it is hard to take decisions regarding research priorities solely on the basis of expectations concerning the best solutions for existential problems.

Second, we can see research as part of realising more complex *social, moral, and cultural projects*. Human beings do not only want to survive and be protected against illness, natural catastrophes, wars, and terrorists. They want to live lives in which they realise specific goals, projects, values, or ideals. For example, we want to live in a democratic society in which the citizens themselves legitimize power. Such a society presupposes that citizens are empowered to form political opinions of their own, are capable of articulating political views and justifying those views in political discourses, and of developing instruments for complex decision-making. Moreover, we want to live in societies that are socially just and inclusive, where people with different mental and bodily capacities and different social and cultural backgrounds can find ways of leading fulfilling lives. We want to have societies that are culturally interesting and diverse, where a variety of cultural and aesthetic forms of expression are possible. For all of these projects to be realised, we need competences from different academic disciplines. Here too we do not know in advance which disciplines will be particularly important for supporting their realisation. Important developments in the history of research have often come from unexpected sides. Moreover, when it comes to cultural and creative dynamics, it is crucial that we remain open to the unplanned.

Third, humans are self-reflexive beings. Whilst engaging in the projects mentioned above, they reflect upon what they are doing and what they are valuing. Humans are driven by the ambition of understanding the natural world, understanding the history of humankind, and understanding the basic conditions of their existence (such as language, culture, behaviour, emotions, and cognition). They want to understand how religion influences their interpretation of social institutions, or how their emotions and cognitions are influenced by biological processes. They want to understand how Chinese art opens up other views than Western art on the interpretation of the world. The drive to understand is not a mere extra (a 'bonus') for a fulfilling life, an indulgence of one's curiosity. It is an integral part of a fulfilling life. Human beings cannot formulate (and reformulate) the projects mentioned above if they do not reflect upon them. They need orientations for practical living. A reflectively oriented life simply is the human way of living. Now, academic research comes into the picture because it can help humans by *enhancing their self-understanding.* Having an adequate understanding of the world around us presupposes research about the phenomena in question, and all academic disciplines, from history to literature studies, from psychology to biology, from medicine to physics, can help in providing this knowledge. The enormous amounts of popular scientific books testify to the fact that humans eagerly absorb academic knowledge.

Moreover, self-understanding also requires a more fundamental understanding of the basis of understanding itself. If we aim at understanding, we must first know what understanding means, and what the presuppositions of understanding are. This requires an understanding of logic, hermeneutics, and ethics, which we know from the history of philosophy, in which people have thought about the possible ways of orienting ourselves in the world, and what reasons we have to assume that some of these interpretations are better than others. It is impossible not to make assumptions about the conditions of understanding. Developing a research question and a research methodology already presupposes that we can give an account of what we want to understand, and in each possible account we already make contestable presuppositions about what it means to understand something. That we make such presuppositions is not a bad thing. We are simply not capable of engaging in any research process if we do not make presuppositions.

However, these presuppositions are always contestable, and this means that research requires self-reflexive activities in which we understand *how research itself is embedded* in the way we orient ourselves in the world. Hence, the self-reflexivity of human life calls for academic research, but this in turn calls for *self-reflexivity within academic research itself*.

To sum up: (i) humans need to live a *physically stable* and where possible *comfortable life*, but (ii) they do so in order to lead a *good life, a meaningful or fulfilling life*, and (iii) they need to be *self-reflexive* about those outlooks on their life. Now, what does all this mean for the organisation of research?

The three aims of research cannot be seen as independent from each other in practice. The first of the three aims outlined above is very often the starting point of research. But at the same time, those research activities can only be understood in the broader context of attempts to develop more complex forms of living and understanding. From a practical perspective, a stable political and economic background is required for being able to commence substantial research activities. Research does not start from nowhere, rather it is normally the case that very concrete problems motivate people to ask more systematic questions. *These considerations make establishing a strict differentiation between 'curiosity-driven' research and research that is 'socially relevant' dubious, for principled reasons.*

On the one hand, research is self-destructive if there is insufficient room for the *internal logic of the research processes* to develop. The principal reason for this has been noted above: there is inherent uncertainty in developing technical solutions as well as social, political, and economic institutions to deal with human problems and aspirations. These uncertainties in 'real life' need to be mirrored in the academic research process, so that the latter is characterized by sufficient flexibility. We cannot predict in advance which scientific solutions and directions will turn out to be promising, and which ones will turn out to be dead ends. This means that our research agenda should not be so fixed as to stifle this internal dynamics.

On the other hand, this 'internal' dynamics does not refer to a process which is contained within the walls of the university. It refers to the self-standing dynamics of society and science co-evolving over time through mutual interactions. There is no reason to defend an ideal of the *uselessness* of research which would only consist of research as a pure, Platonic form of understanding (*theoria,* as the perception of eternal truths). The whole opposition is flawed right from the beginning. We have reasons to see research as instrumental with regard to central projects of human life in general and modern societies in particular. But at the same time, research must be seen as a much broader attempt to get a more reflexive understanding of the world and the self-understanding of human beings. General reflections about the presuppositions of understanding and partial contributions from various research activities to our broader understanding of the world need to be critically discussed within a broader interdisciplinary discourse on the relative merits of each to our understanding of the world. Both elements are crucial for the possibility of fulfilling central research tasks at all.

This leads us to a conclusion that we deliberately want to put in a paradoxical form: *research can only fulfil its instrumental tasks if done in a context in which it is experienced as having the intrinsic value of contributing to human self-understanding.* This paradox points to a tension, but not to a contradiction. It points to an *agenda for organising research agendas:* we need to create research policies in such a way as to honour this paradox in academic practice. With this core message in mind, we now return to our two earlier questions: should research priorities be decided by individual researchers or by a larger community of stakeholders? If it is the latter, is there a role for 'democracy' in setting such priorities?

Science and society: connectivity *and* distance

The proposal of seeing the aims of research as linked to the quest for human self-understanding leads to two seemingly contradictory suggestions for the institutional design of academic research. We can formulate these as the requirements for 'reflective connectivity' and 'reflective distance', respectively.

On the one hand, researchers need to be *reflectively connected* to a wide range of social actors, for the simple reason that it is their human needs, cultural projects, and ultimately their self-understanding that is at stake. If research relates to the aims of human life itself, by helping social actors to deal with the challenges they face, then researchers need to be well-connected to the actors who have these aims. This may have different implications for different disciplines. The medical sciences will require persons who are willing to donate their corpses to scientific research and volunteers to test new medications. Political science will require access to the political process, for example through a willingness from politicians to give them a 'look behind the curtain'. Business studies will require cooperation with businesses to study the outcomes of different business strategies, and so on. Academic researchers function as 'second-order actors' whose activity is connected to the lives of 'first-order actors', be it their physical, economic, or political life. These connections in some cases become structural, and are the basis of co-funded research conducted in cooperation with societal partners. But even if they are not, such linkages are vital to the research process. This also implies that first-order actors may have unique suggestions for research questions. If something is troubling, a cause for wonder, or inspiring for a first-order actor, this is a *prima facie* reason for researchers to consider whether it may lead to new research.

Structurally, the academic landscape in its entirety needs to be arranged in such a way that there is high-quality research on all three levels that we have distinguished, as well as good connections between these levels. The former presupposes that research funds are allocated in a coordinated way, so that no 'gaps' emerge in the research programme that we have outlined. This requirement may conflict with an uncoordinated way of channelling funds to universities, where in practice the number of students in a given discipline determines the amount of research in that discipline. However, it may also conflict with the practice of allowing disciplines to compete for research projects, where this would make it systematically more difficult for some disciplines to get adequate funding for educating new generations of researchers. The latter requirement (connections between the levels) presupposes the organisation of an interdisciplinary dialogue. Many of the cluster questions and routes that the NWA is currently setting up should ideally evolve into platforms for such dialogues.

All in all, we can see that the research programme of enhancing human self-understanding requires a carefully crafted research landscape. It rules out the 'anything goes' policies that are sometimes associated with the idea of individual researchers' academic freedom. However, there is another side to the coin, which we propose to formulate in terms of the requirement of 'reflective distance'.

Researchers need to have *reflective distance* from concrete practices in order to be able to fulfil their tasks. It is the task of companies, policymakers, civil servants etc. to provide solutions to practical problems (whatever they may be). Researchers, as second-order actors, have a different task. They have to be able to distance themselves from the concrete pressures of these practices for various reasons. *First of all*, it is crucial that research is *reliable and independent*, in the sense that it ought not to be seen as a failure if a given research project does not produce the desired results. This is a central insight from the research scandals of recent years: too much pressure on the system makes it very likely that research results will be biased or even corrupted. It is crucial for an open research atmosphere that measures are taken to counterbalance this external pressure. *Secondly*, we want researchers to be *creative* and to develop perspectives on social problems that the practitioners could not develop themselves; otherwise it would make more sense for practitioners to develop these solutions directly. *Thirdly*, researchers are capable of viewing practical problems through *long-term perspectives*. Researchers are good at asking questions that are not present in society, they develop views that are relevant in the long run, and they come up with ideas that will have societal impact in the future. This is a strength that is complementary to that of policymakers who need to come up with quick solutions. Seeing policy problems through long-term perspectives is precisely what is missing with regard to most political problems. Finally, research should not be instrumental with regard to *specific* social groups or specific social ideas. It is rather a characteristic of publically financed research that it is *relevant for society at large* and that it discusses, in an unbiased manner, various views, ideals, and normative starting points of various partners. It is the role of researchers to critically relate to those starting points and not be dependent on normative decisions of societal partners.

These reasons for reflective distance provide a good reason for the independence of the academic *community* as a whole from social actors who would want to influence the research priorities of universities. However, at least to a certain extent, they also provide a good reason for giving *individual* researchers sufficient room for manoeuvre, since innovative strategies for reliability, creativity, and long-termism must, in the final instance, come from them. The question then becomes how the academic community can organise its own research agenda, so that it (i) does justice to its central mission of helping social actors enhance their self-understanding, and (ii) provides individual researchers with sufficient flexibility to contribute to this task.

Taking up this challenge is made more difficult by the *inner dynamics of research* itself. In recent decades, research has become more diversified and specialized. Over the course of the 20th century, something akin to a generally accepted canon of relevant disciplines, or a hierarchy of accepted research topics, became increasingly contested. There no longer is a shared research culture or generally shared philosophical understanding of what good research is, nor of which methodologies or epistemic standards are appropriate. There exists a kind of local consensus within limited research communities, but there is no broadly shared understanding of basic assumptions regarding research. This implies that discussions about research priorities are a matter of dispute that cannot be settled by reference to tradition and consensus. Moreover, specialization has reached such a level that it is hardly possible for researchers to *oversee* broader fields in general.

The generalist that has informed opinions about the most relevant developments in all research fields becomes the exception, most researchers are glad if they are able to well-informed on the newest developments in their own field. Finally, due to *internationalization* the context of research is in fact the whole world. In some research fields people are participating in cooperative research that is virtually active on a global scale. Coordination of research on a global scale, however, only happens within some selected fields; there is no structured dialogue, rather than incidental dialogue, about research priorities.

All of these factors make priority setting in research necessarily difficult. Researchers are not sufficiently well organised to deal with disagreements about research priorities in a rational way. There is a serious lack of fora for rational discourse about the integration of specialist research into broader perspectives. Discourses about rational reasons for research to develop in one direction instead of another are hardly possible if researchers are unable to oversee the relevant fields, and are at the same time always competing for research money. That research is developing in specific directions for rational reasons is the basis for the trustworthiness of academic research (and so, in the final instance, for stability in public funding of such research). The only forum where researchers make decisions about priority setting in research is within various research organisations in decisions about research projects, and in the selection of new researchers within universities. The logic of these decisions is, however, primarily concerned with the quality of the researcher, and not with the relevance of his research topics. In practice, it will not always be easy to distinguish both aspects, but in any case these decisions do not aim to decide research priorities on the basis of a systematic analysis of research desiderata, research needs, and a systematic process of weighing arguments for or against different possible research agendas. How can we do better?

Organising deliberation about research priorities

Given the need for a research process characterized by both reflective connectivity and reflective distance, we want to highlight four aspects that we think to be crucial for future debates about priority setting against the background of the NWA:

First, it is not self-evident that research priorities should be set on a *national level*. In fact, the NWA already has several competitors. For example, the EU has an elaborate programme of research funding, addressing the themes that EU civil servants have drafted in conjunction with scientists. Many universities have research focus areas of their own, into which they channel funds. Research schools (stretching over various universities), private foundations sponsoring research and others are trying to influence priorities. In this landscape, it should be made clear whether the NWA is just one amongst many initiatives, or if it somehow fulfils a coordinating role. It is not very efficient, to say the least, to have the same conversations about research priorities at different levels.

Second, the process of the NWA also forces us to reflect on what it means to formulate good research questions. Research questions are not simply formulated to increase knowledge in isolated topics. Research questions should rather be directed towards the increase of understanding of a specific domain against the background of an understanding of this domain in a broader sense. We do not simply investigate specific genes; we rather develop an understanding of the functions of genes within broader theoretical outlooks on the human body, nature, and life. This implies that the development of research questions should always be embedded within broader interdisciplinary discourses about these theoretical perspectives. The fora for those discourses are highly underdeveloped; national research schools or interdisciplinary centres and the like could fulfil crucial roles in the development of new research perspectives. What we need is for researchers to have a much more active role in the development of in-depth analyses about long-term research strategies, rather than selections of research questions on specific topics. This would probably presuppose new fora for these analyses. In any case, it is important for researchers to better organise themselves to be able to play a more active role in the setting of priorities.

Third, given the demand for connectivity, it is essential that there is *input from the public*. However, the way this was done in the NWA was rather coarse-grained. Simply asking everyone to deliver research questions leads to a process which not only generates too many questions but is also subject to manipulation (as researchers were themselves very active in submitting research questions) and it leads to many questions which are either already answered or unanswerable. In the light of our last point, it would be necessary to embed the formulation of a research question into broader theoretical discourses in order to understand how this input relates to our current understanding of those research domains. The challenge for the future is to involve citizens in a more constructive way. This will probably require more organised contacts with various social practices in which people struggle with physical, cultural, economic, and other challenges. Many contacts already exist; the challenge is to bring this to bear on research priority-setting in a transparent way.

Fourth, there is a *problem of representativeness* in the current setup of the research agenda. Apart from academic institutions, only organisations of employers (VNO and MKB) are partners in the knowledge coalition. This reinforces existing impressions that economic interests (and only those on one side of the field) are societal partners with a stake in research. Given the centrality of human self-understanding, a much broader coalition is necessary, from Amnesty International to Urgenda, from consumer organisations to the World Nature Fund. We could even make a case that political parties, as important representatives of practical views on the future of society, should be involved. However, instead of putting all of these partners in a 'steering position' (which is what the knowledge coalition does for some representatives of industry), it would be better if they were conversation partners at a functional distance from the academic community, which itself has a steering role.

Fifth, we can wonder what instruments are appropriate for engaging with the representatives of various organisations. It seems probable that the method of co-financing in specific projects is not the most productive instrument. It is much more important to establish settings where leading figures from these organisations discuss the long-term perspectives of research together with interdisciplinary groups of researchers. Joint projects make sense only if they are embedded in a more in-depth understanding of long-term perspectives. These discourses should, however, really be for analysis, and not simply brainstorming meetings, as they have been in the case of the NWA.

Sixth, a crucial question is to what extent there could be a role for democracy in this process of priority setting. It seems that Parliament and government primarily play a role when it comes to research policy if direct economic interests are at stake, or when it comes to research that is directly relevant for specific policy areas. There is, however, hardly any serious involvement of democratic institutions in the development of research priorities. It is evident that there are limits to the meaningful involvement of political institutions, not only because of questions of competence, but also for more principled reasons – after all, in the past the fight for academic freedom was one against the direct intervention of public authorities in the independence of the universities. However, if industry has an important impact in the development of research, it is strange if democratically legitimized institutions do not. This indicates that a new relationship between the roles of political institutions, societal interest groups, and the researchers themselves needs to be found – a new relationship that reflects the public role of universities, and at the same time ensures the academic freedom that researchers need to fulfil their public task.

Seventh, an agenda presupposes a *fixed menu* of items. There needs to be a more developed idea of how this impacts upon a scientific process that itself rapidly changes every day. There is a risk of 'shooting at a moving target'. If the agenda is to become a well-established part of the Dutch research process, procedures for revision and updating need to be made so that the agenda does not fossilize into an overview of yesterday's priorities.

We hope that these suggestions provide constructive proposals for reflecting upon the current process of establishing a National Research Agenda.

No University without Diversity

The Dynamic Ecosystem of Scientific and Social Innovation

André Knottnerus

Introduction: the Dutch National Research Agenda

In 2015, a public and media-supported invitation to the Dutch community to submit questions to science was the start of a process to develop a national research agenda. The chosen approach was bold and innovative and drew a lot of attention, nationally and internationally. As has been described elsewhere in this volume, it turned out to be a successful approach, harvesting close to 12,000 questions from a broad range of highly motivated individuals and groups with various social backgrounds – citizens, consumers, professionals, businesses, policymakers – and also from researchers from the full spectrum of scientific disciplines.

After review, these specific questions were clustered in 140 more general questions covering key fields of scientific, social, and economic interest and related to existing institutional research and knowledge agendas and the EU grand challenges. In 2016, the implementation process started, and work is being done to further connect bottom-up initiatives with identified challenges and to make investment plans to facilitate corresponding research programmes.

A central aim of the Dutch National Research Agenda is to stimulate scientific creativity in the broadest sense and to harness this creativity to meet important scientific and societal challenges. The objective is to establish an adaptive, resilient, and dynamic research system that on the one hand is connected to science-driven research agendas but at the same time can be sensitive and responsive to important developments in society.

In this essay I shall explore how characteristics of the Dutch National Research Agenda relate to international developments taking place in the world of research and innovation. In doing so I shall use the agenda as cause and point of reference for some reflections on how to create a more dynamic ecosystem for science and innovation, allowing me to assess the relevance and timeliness of the agenda.

The importance of diversity and connectivity

To be effective and productive, given the fast and complex scientific and societal developments, a research agenda must have connective power and be able to promote collaboration across disciplines and sectors. In this context, the following insights are paramount.

New ideas and insights emerge best in a multiform and diverse multidisciplinary landscape in social interaction with external peers (Blackwell et al., 2010; Hasan and Koning, 2015a; Hasan and Koning, 2015b). Original and unexpected approaches and solutions of a problem have a higher chance to occur at the interplay of different disciplines which might not even have collaborated before on that problem. Such "bottom-up" creativity is facilitated in a scientific community bringing together a broad range of expertise. At the same time, "top-down" approaches to address grand scientific and societal challenges in innovative ways cannot be achieved without a broad input of creativity. In other words, also for this purpose diversity and the presence of multiple disciplines and research approaches are a precondition.

Moreover, innovation is facilitated by opportunities for direct face-to-face communication. Even in an era of increasing internet connectivity and new possibilities for real-time worldwide communication, 'physical connectivity' within short geographical distance remains important. It is a key requirement for creative brainstorming and serendipity, knowledge circulation, and productive collaboration (WRR, 2013; Rosenman, 2001). For this reason, processes of innovation are accelerated in urban regions ('smart cities') and advanced universities comprising a comprehensive diversity or 'multiversity' of disciplines and talents (NAS, 2012; Ter Weel et al., 2010).

Successfully realising innovative ideas at the interplay of different disciplines – i.e. interdisciplinary research – requires specific conditions to be met, both in research and research policy (Grensverleggend, 2015). It requires bridging divides not only between top-down and bottom-up research programming and between untied research and research focusing on societal challenges, but also between public and private funding of research, and between the academic world and the general public. This can be achieved if we leave the world of walls and separations and agree to be players in one dynamic ecosystem of research and innovation.

The formulation of the Dutch National Research Agenda fits these patterns and responds to this emerging ecosystem in that it involves the public and societal actors in setting the agenda and connects players across the entire knowledge and innovation landscape.

Integrating top-down and bottom-up approaches

Building on the previous section, we must recognise that in discussions on how research and innovation should be promoted, there are two perspectives that are both old and still very much alive. First, there is the perspective of the autonomy and freedom of research, as a *conditio sine qua non* for continuous progress and development, especially in the long term. This is also the perspective of bottom-up initiatives, science-driven governance, and self-regulation of research, leading to a multiform spectrum of largely unpredictable scientific creativity.

The second perspective is the one of research being motivated or stimulated by policies to meet grand challenges for science and society. In the context of this second perspective, clustering of research capacity in larger themes, programmes, or centres has become common practice in order to combine critical mass with focus to address the defined challenges with available resources.

While each of these two perspectives represents a strong case in itself, it is a continuous challenge to connect them in one and the same comprehensive research strategy. While there are still researchers and policymakers who believe that these perspectives are incompatible, the Dutch National Research Agenda seeks to overcome this distinction by presenting a framework that allows room for both approaches and integrates them. In doing so it builds on an international trend of linking bottom-up creativity with top-down programming. Examples of this trend are worldwide scientific collaborations such as the Human Genome Project and regional geographic concentrations to serve scientific and economic progress. The latter has been demonstrated in the NASA space research programmes, the CERN collaboration in Geneva, Silicon Valley in California, and the Eindhoven Brainport area. These successes were by and large the result of combining public and private efforts from academia, government, and industry. Public and private initiatives and extensive funding programmes – integrating basic, strategic, applied, and practice-related expertise – have also enabled Wageningen University to be a world leader in food and malaria research. In all these cases, concerted collaboration guided by common goals has shown to be a determinant of creative and interdisciplinary development.

The Dutch policy to promote economic top sectors is another case in point. In these top sectors, which are both publicly and privately financed, government, science, and industry work together to create innovative products and services to address major societal challenges and to increase the earning power of the Dutch economy (Adviesraad voor Wetenschap, 2014). The Dutch National Research Agenda embraces and builds on the experience of this approach, but is wider in scope, seeking a more balanced representation of input from natural and life sciences, social sciences and the humanities, and gives more space to bottom-up initiatives. Moreover, it more explicitly emphasizes the need to add a diversity of other societal actors to the research collaboration between government, science, and industry.

This development is indicative of a wider trend in which the distinction between the public and private domain is changing, as considered by the Netherlands Scientific Council for Government Policy (WRR, 2013). Whether it comes to energy conservation, sustainable production, the implementation of a basic package of healthcare services, privacy, security, or reliability of financial markets, governments cannot serve the public interests without a strong commitment from the private domain, both nationally and internationally. At the same time the private sector is increasingly dependent on cross-border public policies. Accordingly, a public-private continuum has emerged, where players pursue their goals in mutual interconnection. In fact, one could speak about a repositioning of both domains in a context of shared responsibility, according to a broader conception of the public interest as already conceived by Spinoza (2010), cause' which is relevant for and must be supported by all of us together. It is precisely this interpretation that fits with the challenges of today and tomorrow, also in the field of research policies and research funding.

No more hierarchy of sciences

The trend to merge top-down programming with bottom-up creativity heralds the end of hierarchical dividing lines between the sciences. Equity and mutual respect in the appreciation of and between the various sciences is a *conditio sine qua non* for productive and motivating interdisciplinary collaboration.

Illustrative in this context is the anecdotic 'Feinstein cycle', presented by Alvan R. Feinstein, founder of modern clinical epidemiology, at a seminar on clinical epidemiology and healthcare research at Maastricht University in 1989 (Knottnerus, 2012). A biologist rather condescendingly tells a biomedical researcher that the latter merely applies the knowledge of mechanisms detected by biology. Next, a chemist challenges the biologist, claiming that not biology but chemistry has provided the fundamentals of these mechanisms. No, says the physicist, our laws of nature and elementary particles determine all of the living and the non-living world. Subsequently the mathematician poses that physical empiricism merely confirms what he and his colleagues had predicted. The philosopher pushes him aside by saying that math is just one instrument of fundamental thinking, not necessarily the most illuminating. Then, in the retake, the biomedical researcher takes his second chance: 'Colleagues, you have shown that human thinking needs improvement and it is up to my field to enhance brain performance.'

The obvious catching point is that, instead of claiming that any discipline is more basic than any other, it is more fruitful to accept that all disciplines need one another, and have their own unique value as part of the big mosaic of sciences. This certainly is the spirit behind the Dutch National Research Agenda, which has identified research themes that require concerted efforts of many different disciplines.

For the various disciplines to better understand and appreciate one another, there is also a need for a better system of quality assessment of scientific performance across disciplines. Quality assessment should take the diversity of sciences as a starting point and be sensitive to differences in publication and citation cultures. Moreover, various types of scientific and social impact should be taken into account (Knottnerus et al., 2002; Knottnerus, 1988; Wetenschapsvisie 2025, 2014).

There is growing criticism that quality assessment of research has become 'quantity assessment' focusing on counting publications and citations. A fair and more scientifically acceptable assessment can be facilitated by (re)introducing quality of content review: by looking at what in fact has been accomplished; reading, not just counting what has been reported; reviewing originality, quality, and contributions to real progress; and also being critical as to wasting of resources and efforts, and unnecessarily burdening of study subjects and guinea pigs (MacLeod et al., 2014). The use of assessment criteria such as contributing to progress and appropriate use of resources is especially essential at the level of programme clusters and institutions. Moreover, at that level respecting differences in publication culture is extra relevant since in a cluster context comparisons between disciplines are more directly made.

It would not be surprising if such a change in orientation of the assessment system would lead to different quality rankings of researchers and institutions. Originality and innovation, and promising long-term impact – elements that are not easily recognized in a system mainly focusing on past performance and recent citations – would be earlier detected, published, acknowledged, and stimulated.

Public involvement as a game changer

Giving the general public a significant role in developing the Dutch National Research Agenda was seen by many as a bold, innovative, and risky step. Yet it is indicative of a wider trend of growing public involvement in science. The once sharp demarcation between scientific authority and the world of the layman (who was expected to just accept and respect that authority and to unconditionally pay its costs) is rapidly fading. This is a result of a general development that also has a major influence on the societal position of research and on research policy: the continuously increasing commitment of the public (Gregory and Miller, 1998).

While in the sixties research policy was merely a matter of interaction between governments and research institutions, over the past twenty years the perceptions, opinions, and involvement of the public have become a decisive factor. This has been strongly facilitated by the rising levels of education and social emancipation of western populations and by the media. At the same time, the research community has recognized much more than before that the public's trust and confidence in science is crucial for social investments in research and innovation.

This is associated with more direct public accountability: research institutions and groups, but also individual researchers, are now challenged to explain to the public the work they do, why this is important and useful, and why their work needs and deserves public investments. Where in the past politicians could annually decide on those investments in a 'backstage context', nowadays such decisions need explicit public support. Not only since politicians and their decision-making are much more under day-to-day public pressure, but also because in the political arena the various priorities are more transparently brought into intra- and inter-sector competition.

Some consider this enhanced involvement of the general public to pose a risk, in the sense that investments in research and development may become susceptible to short-term fluctuations evoked by opportunism. But it can also be seen as an opportunity to build a direct, strong, and stable mutual relationship with the public, with a view on the longer term future and to provide safeguards against political 'short-termism'. Indeed, it is an important task of responsible stakeholders to provide well-balanced countervailing information, that is, checks and balances based on which the public and politicians can develop long-term views. It is precisely for this reason that the scientific community must actively connect with the public to ensure the indispensable societal foundations for the future of sciences.

As the future of research is increasingly dependent of the public's confidence and trust, public accountability no longer allows 'scientific isolationism'. Public accountability is accompanied with public involvement, and is a sine qua non for sustainable investments in research.

The role of government

The Dutch National Research Agenda was developed at the request of the Dutch government. Rather than formulating such an agenda itself, the government judged it better to ask the main players in the Dutch knowledge and innovation system to develop the agenda through a participatory approach allowing for public involvement.

This makes us reflect on what role government can have in promoting the result: a both scientifically and socially relevant research agenda. In a globalizing world where international and European research agendas play an increasing role and in which private actors have a major, often transnational impact on research and innovation, national governments by themselves are less powerful than before. But they can stimulate fruitful conditions, interacting with science, industries, and social organisations. If convincingly done, this can leverage much greater public impact than would otherwise be possible. In this respect, the initiative of the Dutch government to invite a steering committee with participation from the public and private sectors to develop the national agenda was well considered.

But as research and innovation are internationally competitive activities, public-private cooperation must not be seen as a means to cut down public research funding, certainly not in a situation in which the Dutch investments in research and innovation are lower than those of many comparable Western countries (Deuten, 2015). If we want to optimally utilize and maintain the high 'specific gravity' and comparatively excellent performance of Dutch research and innovation (Prestaties in perspectief, 2012; Cornell University et al., 2015; BiGGAR report, 2010), the jointly deployed relative volume of research resources should at least keep up with international trends. Today's good performance reflects the impact of investments of many years ago, not just of today or yesterday. This is a crucial issue for society's resilience in an uncertain future and therefore represents a major mission for the Dutch government. Being able to maintain a solid base of knowledge- and curiosity-driven research is vital as a foundation for problem-driven research and longer term social gains (Ruimte voor ongebonden onderzoek, 2015).

A related government responsibility is to monitor and safeguard diversity and vibrancy of the research and innovation ecosystem in the interest of long-term resilience. This also implies allowing sufficient space for research that cannot easily be translated into social or economic value in the foreseeable future.

Obviously, the public interest of huge parts of research is that it may result in – often unforeseen – long-term applicability for the public good, and that it is rooted in mankind's proven conviction that we can only be what we are if we continue to look for not yet understood pasts and unknown futures. Think of climate research, basic molecular biological and psychological research, and historical and philosophical research.

One may also think of research topics like safeguarding human rights in deprived areas, or evidence-based discontinuation of excessive, long-term multiple drug use (Centrum voor ethiek en gezondheid, 2009). These topics are obviously crucial for society but not immediately attractive to the market. Consider as well long-term investments in research infrastructures that will not be achieved by single private parties because of market uncertainties, but are nonetheless essential for future generations not yet sitting at the stakeholders' table.

In this context, given limited public resources, a logical question is whether economically attractive research should not be more extensively financed or refunded from the benefiting markets, so that more vulnerable research activities could be better safeguarded by public funding. This would protect the latter against 'market failure', and would also facilitate the 'incubator' and 'back-to-the-drawing-board' functions of academia, which are, in the end, in everyone's interest.

Finally, in the currently complicated geopolitical context, with its increased emphasis on national interests, international scientific and expertise-based cooperation should not be pushed into the background. Such cooperation is both natural and essential for science itself, which needs thinking in and exploring of a world without borders. Moreover, international and especially European scientific cooperation is a prerequisite for addressing cross-border issues such as environmental quality, building and utilizing expensive infrastructures, and optimally handling rare diseases (Knottnerus, 2008), but also for effective competitiveness in a world with increasingly large players (WRR, 2010). It is therefore a good thing that independent researchers at the scientific workplace are keeping a cool head, irrespective of all tensions in the political arena, and continue building international bridges and breaking down walls.

Conclusion

After this short *tour d'horizon* of national and international developments in science and innovation, we are in a better position to situate the Dutch National Research Agenda in the context of emerging patterns. We have seen that the agenda reflects, responds to, and brings to the fore wider trends related to research and research policy.

The chosen approach is timely and meets the needs of our time, and is therefore very promising. Optimism is also warranted as it is a politically adopted innovative approach, and as stakeholders are creatively working together to overcome any hampering dividing lines, such as those between the scientific and the 'lay community', between public and private actors, and between the natural sciences, the social sciences, and the humanities.

The agenda is bound to play a role in the newly developing ecosystem of research and innovation, which is to be flexible and tailored according to the expertise and commitments needed to address major scientific and societal challenges.

References

Adviesraad voor Wetenschap, *Technologie en Innovatie. Balans van de topsectoren* (The Hague: AWTI, 2014)

Blackwell A., L. Wilson, C. Boulton and J. Knell, 'Creating value across boundaries. Maximising the return from interdisciplinary innovation', *UK National Endowment for Science, Technology and the Arts,* Research report: May 2010

Centrum voor ethiek en gezondheid, 'Wie betaalt, bepaalt? Over financiering en het ontwikkelen van medische kennis' (The Hague: Gezondheidsraad/CEG, 2009)

Cornell University, INSEAD, and WIPO, *The Global Innovation Index 2015: Effective Innovation Policies for Development* (Fontainebleau/Ithaca/Geneva: The World Intellectual Property Organization, 2015)

Deuten, J., *R&D goes global: Policy implications for the Netherlands as a knowledge region in a global perspective* (The Hague: Rathenau Instituut, 2015)

BiGGAR Economics, *Economic Impact of Research & Commercialisation at Leiden University & Leiden University Medical Centre. A Report to Leiden University Research and Innovation Services* (Roslin, Scotland: BiGGAR Economics, 2011)

Gregory, J., and S. Miller, *Science and the public. Communication, culture, and credibility* (Cambridge: Basic Books, 1998)

Grensverleggend, *Kansen en belemmeringen voor interdisciplinair onderzoek* (Amsterdam: De Jonge Akademie, 2015)

Hasan, S., and R. Koning, 'Conversational Peers and Idea Generation: Evidence from a Field Experiment', Stanford GSB, September 29, 2015(b)

Hasan, S., and R. Koning, 'Conversational Peers, Team Dynamics and the Ideation Processes: Evidence from a Randomized Field Experiment', Stanford University, August 18, 2015(a)

Knottnerus, J.A., 'Improving Health and Health Care across Europe: An Integrated Approach', in *Health Care Policy and Fundamental Rights in Europe,* edited by B.R. Machiavelli and F. Velo (Rome: European Liberal Forum, 2008), pp. 51-62

Knottnerus, J.A., 'Wetenschap en de publieke zaak (Science and the public good)', Keynote lecture at the opening of the Utrecht University Academic Year, 3 September 2012

Knottnerus, J.A., 'Dialectiek van het onderzoek in de huisartsgeneeskunde' Maastricht: Rijksuniversiteit Limburg, 1988)

Knottnerus, J.A., J.M. Bensing, L.M. Bouter, et al., *The Societal Impact of Applied Health Research: Towards a Quality Assessment System* (Amsterdam: Royal Netherlands Academy of Arts and Sciences, Council for Medical Sciences, 2002)

MacLeod, M.R., S. Michie, I. Roberts, I. Chalmers, U. Dirnagl, J.P.A. Ioannidis, Salman R. Al-Shadi, A.-W. Chan, and P. Glasziou, 'Biomedical research: increasing value, reducing waste', *The Lancet,* 383, 2014, pp. 101-104

National Academy of Sciences, *Research Universities and the Future of America: Ten Breakthrough Actions Vital to Our Nation's Prosperity and Security* (Washington, DC: The National Academy of Sciences Press, 2012)

'Prestaties in perspectief. Trendrapportage universiteiten 2000-2020' (The Hague: VSNU, 2012)

Rosenman, M.F., 'Serendipity and Scientific Discovery. Creativity and Leadership in the 21st Century Firm', Volume 13, 2001, pp. 187-193

Ruimte voor ongebonden onderzoek. Signalen uit de Nederlandse Wetenschap (Amsterdam: Koninklijke Academie van Wetenschappen, 2015)

Spinoza, B., *Tractatus Theologico-politicus* (1670) (Klever W. Delft: Eburon, 2010)

Weel, B. ter, A. van der Horst, and G. Gelauff, *The Netherlands of 2040* (The Hague: CPB Netherlands Bureau for Economic Policy Analysis, 2010)

'Wetenschapsvisie 2025, keuzes voor de toekomst' (The Hague: Ministerie van Onderwijs, Cultuur en Wetenschap, 2014)

WRR (Scientific Council for Government Policy [Wetenschappelijke Raad voor het Regeringsbeleid]), Aan het buitenland gehecht. Over verankering en strategie van Nederlands buitenlandbeleid (Amsterdam: Amsterdam University Press, 2010). English translation: *Attached to the World. On the Anchoring and Strategy of Dutch Foreign Policy* (Amsterdam: Amsterdam University Press, 2011)

WRR (Scientific Council for Government Policy) *Naar een lerende economie. Investeren in het verdienvermogen van Nederland,* WRR-rapport 90 (Amsterdam: Amsterdam University Press, 2013). Synopsis in English: Towards a learning economy. http://www.wrr.nl/fileadmin/en/publicaties/PDF-samenvattingen/Summary_Towards_a_learning_economy.pdf

WRR (Scientific Council for Government Policy), *Publieke zaken in de marktsamenleving* (Amsterdam: Amsterdam University Press, 2013)

Inspiration

Louise Gunning-Schepers

A National Research Agenda

When the Knowledge Coalition set out to develop the Dutch National Research Agenda, their mission was to come up with an inspiring product. Looking back, the process – maybe even more than the product – did indeed prove to be inspiring. Many more individuals than expected, both researchers and individual citizens of all age groups, submitted a question that they would like to answer or see answered. And many more individuals attended the conferences, the festivals, and the 'living room lectures' that were organised, and watched the debates about the questions on television. These individuals personified the inherent curiosity for new knowledge that drives research and innovation, but that apparently also inspires society.

The time and effort that many academics, especially from the Young Academy, put into clustering the almost 12.000 questions into the final 140 overarching questions that make up the Dutch National Research Agenda went beyond the call of duty. The questions cover all fields of science (in the Dutch sense of the word), all disciplines and all stages of research and innovation. The questions also build on and connect the research agendas of the different partners of the Knowledge Coalition. In this way the Dutch National Research Agenda inspired these different partners (Universities, University Medical Centres, Universities of Applied Sciences, the Royal Netherlands Academy of Arts and Sciences, the Netherlands Organisation for Scientific Research, the Applied Research Institutes, Industry and Small Scale Enterprises) to step up collaboration and jointly shoulder the responsibility to implement the agenda.

Not only the process, but also the product itself should serve as a source of inspiration, otherwise the mobilization of so many citizens and busy researchers would be in vain. This product, the 140 questions in the agenda, will be put to the test in the coming year. Will it help secondary school students to choose a career in a certain field? Will it inspire students to write their thesis on a topic that links to one of the agenda questions? Will academics from different disciplines find one another, looking at the same questions from different perspectives? Will the perpetual exchange between basic research, applied science, and implementation accelerate innovative solutions in society and business?

To help this process we have made the Dutch National Research Agenda available to every secondary school in the Netherlands, with instructions for teachers on how to use it. Through the landscape of the 140 questions we have mapped out 16 routes, thus creating subsets of questions touching upon complex challenges. For each of these routes a number of workshops are being organised where researchers from all disciplines active in these fields and those who use their results will meet and get to know one another and discover what each one is doing. Hopefully these discussions will be so inspiring that new research projects will arise. Research projects in which individuals or disciplines that have not collaborated before find common ground for accelerating their quest for answers. These 'game changers' will look for and find funding to achieve their potential. It is to be hoped that some of these newly formed communities of researchers will continue to meet in the coming years and that additional coalitions of researchers will explore alternative routes. In this way the Dutch National Research Agenda will be able to continually inspire innovative research for years to come.

In the meantime, the Knowledge Coalition will put forward a strong plea for renewed investments in research and innovation by subsequent governments, as it is convinced that research and innovation are the most powerful source for creating growth and jobs, finding solutions for social dilemmas, and fuelling inspiration and ambition for new generations. Substantial and structural additional investments are required, half of which need to be spent on maintaining a strong, broad base for science, in all disciplines and at all levels (including investments in infrastructure and young talent), and the other half to be invested in game changers that have been identified in the route workshops. Why are both needed?

An investment in the future

Academics are part of a global community. Researchers know their peers abroad as well as those in their own country and students travel the globe to find the education they aspire to. This has been the case since Erasmus travelled Europe. In these international communities researchers are familiar with one another's work and know what the truly important hypotheses are. They are often as curious about the results of others as they are impatient to know what comes out of their own experiments. After all these are the building blocks that they will need and use for their next projects.

But only the best are part of the inner circle of that community, where the real new insights are developed and shared. Being part of that much smaller community gives them access to new information long before it is published, information that may be crucial for a new research project or for an application or the development of a new product. Any country that aspires to use science for innovation will want to have access to a large number of such inner circles as it is often very difficult to predict where the interesting opportunities will emerge.

In the Netherlands we are lucky to have a large, diverse, vibrant and very successful knowledge and innovation community. Many leading researchers are part of the inner circle in their field. But having access to the latest information and doing groundbreaking research is not enough. One has to be able to link that to those who can take further steps – in applied research, product innovation, or social innovation. It is this ecosystem that the Knowledge Coalition managed to mobilize for the Dutch National Research Agenda. When the Knowledge Coalition was unexpectedly flooded by a tsunami of questions, many researchers were able and willing to help sort them and cluster them around research communities that were already functioning in such an international context. The 140 questions that make up the current Agenda can thus be seen as the result of connecting citizens who believe science can help us forward with established research networks that can link these questions to the global academic and business R&D community.

Past investments in research in the Netherlands have established a high-level platform supported by many pillars of disciplinary excellence that each have access to the highest level of international knowledge and scientific debate. It is this broad and high-level base that provides the best possible starting position for researchers striving to reach for the very top, be it in basic or applied research or in using that knowledge for breakthrough innovations in societal or economic terms. Since we cannot easily predict where the important opportunities will arise, the academic community in the Netherlands has always set its aim on excellence rather than deciding upfront in which topics to invest. Excellence depends on recognizing, recruiting, and training talent. That is what universities are for. But it is the students who decide to enter a university and choose the subjects they feel inspired by. They will determine in which fields excellence will thrive. That is why half of the investments should be spent on keeping the base broad and strong, but above all attractive so we can keep young talent in the Netherlands.

At the same time, government may choose to invest in certain pressing societal issues, be it for economic growth, jobs, or the well-being of citizens. Their choice may depend on the political constellation of the day or on the opportunities that are perceived. With this National Research Agenda we have made visible pressing issues that require new knowledge and the variety of potential game changers that research and innovation have to offer. That is what the other half of the investments should be spent on and the Knowledge Coalition will present possibilities to do so and mechanisms to identify the most promising initiatives. In doing this we provide politicians with an investment agenda that can help prepare our society for the future. We hope it will prove to be an 'offer you can't refuse'.

Process of Developing the Dutch National Research Agenda

Background

In November 2014, the Dutch cabinet submitted the policy paper 'Vision for Science 2025' to Parliament. As the title indicates, the paper unfolded a vision of the future of Dutch science. It formulated a number of policy ambitions, the most important being that in 2025 Dutch science should hold a top position in global rankings.

The main strategy to realise this ambition was to enhance coherence and impact by a joining of forces. And the central instrument to make this happen was the development and formulation of a National Research Agenda. This agenda was to meet quite some expectations:

> The National Science Agenda will appeal to the imagination; it will inspire and challenge both the research field and society itself to achieve momentous breakthroughs. It will create a better match between research on the one hand, and social and economic needs and opportunities on the other. It will clearly set out those areas in which the Netherlands is to stand out through truly excellent research. By raising the profile of Dutch science with its own agenda, we shall strengthen our position within international partnerships. In specific areas, the Netherlands will take the lead in those partnerships. This is important if we are to attract top talent and safeguard the interests of our knowledge-intensive industry. (Ministry of Education, Culture and Science of the Government of the Netherlands, *2025 Vision for Science: choices for the future*, p. 24)

The aims of the Dutch National Research Agenda were summarised in the mandate letter of 25 November 2014, which stated that the Agenda should:
– identify social themes and top scientific fields;
– build on existing agendas and make connections;
– influence future planning;
– improve the international position of Dutch science and society's engagement in research;
– encourage cooperation and increase its impact throughout the knowledge chain;

– focus on research in which a national approach offers greater value and contributes something that isolated institutions or existing alliances have so far failed to achieve (principle of subsidiarity).

The mandate letter concluded by stating: 'Every matter included in the National Research Agenda should be important, but not every important matter should be included in the National Research Agenda.'

Governance

The mandate to draw up a national research agenda was assigned to the Knowledge Coalition, consisting of the most important umbrella organisations of the Dutch knowledge and innovation system. The Knowledge Coalition installed a Steering Committee responsible for developing the Dutch National Research Agenda. On 23 January 2015, shortly after the mandate had been assigned, the ministers appointed Prof Beatrice de Graaf and Prof Alexander Rinnooy Kan as independent co-chairpersons, deeming them capable of providing authoritative, unifying, and innovative leadership within the process at hand.

The decision to appoint co-chairpersons allowed the burden of work to be shared, reduced vulnerability in the event of absence, and brought different backgrounds and areas of expertise into the process. It also made it possible to benefit from the differences between the two appointees in terms of gender, age, and disciplinary background.

The Steering Committee and chairpersons met once every three weeks from February to December 2015. To ensure continuity and communication with the participating institutions, these meetings were also attended by the official deputies of the Steering Committee members.

The members of the Knowledge Coalition were all part of the Dutch science system and were expected to bear the primary responsibility for implementing the Dutch National Research Agenda. As such, the Steering Committee was considered to be insufficiently representative of society at large. Since it was deemed undesirable to add governmental and civil society parties to the Steering Committee, it was decided to set up a Liaison Group as a separate body. The Liaison Group was appointed in April 2015. Although acting in a private capacity, its members represented a wide range of different social sectors. The Liaison Group offered the Steering Committee solicited and unsolicited advice, attended the preparatory conferences, and built relationships with strategic agendas, knowledge-based institutions, and advisory bodies.

The chairpersons and the Steering Committee were assisted by a secretariat headed by the Steering Committee secretary. The secretariat's staff members were nominated by the members of the Knowledge Coalition. Most were affiliated with the Netherlands Organisation for Scientific Research (NWO) in The Hague, which hosted the secretariat. Some staff members were communication specialists. The secretariat also established ongoing working relationships with the communication managers of the Knowledge Coalition members.

The relevant ministers and state secretary, the chairpersons, and the Steering Committee met every quarter to discuss progress. Preparations for these meetings were undertaken by the directors of the relevant ministries, who coordinated with the Dutch National Research Agenda secretary. The secretary also met with ministerial officials every other week.

Communication

One of the critical success factors for the Dutch National Research Agenda was to ensure a broad base of support among the parties involved and their member organisations. The process of developing the Agenda also gave the participants a unique opportunity to show what Dutch research had to offer and, in doing so, to generate and boost support for science and, more specifically, for the Agenda itself. Generating that support was the focus of the relevant communication activities.

With so many parties involved in developing the Agenda, uniform and consistent positioning was very important. The core communication messages were:
- The Dutch National Research Agenda connects: it builds bridges between existing agendas and unites disparate parties.
- The focus is on the content, and not the financial consequences.
- The Agenda encompasses every type of research, from basic to applied and practice-based.
- The Agenda is inspiring and shows the imaginative power of science.
- The Agenda shows that science belongs to everyone.

The communication activities focused on roughly the following three themes:

1 Creating and maintaining support for and commitment to the Agenda by the parties involved
The website was the main communication platform. It was considered important to allow all parties involved to track the process closely on the website. Partners' communication channels, including social media, were also used.

2 Generating broad support for the Agenda and for science in general
The main channels of communication here were the website and social media, alliances with such partners as *New Scientist* magazine, the Lowlands organisation (a music festival) and Kennislink (a popular science website), as well as a media partnership with the popular television talk show *DWDD* (*De Wereld Draait Door*).

3 Communication as part of the public consultation procedure
The process of developing the Dutch National Research Agenda was demand-driven. This basic premise offered numerous opportunities to express the connective power of the Agenda, something that called for meticulous, transparent and, above all, interactive communication with those who had submitted questions and other interested parties..

Developing the Dutch National Research Agenda

The process of developing and formulating the Dutch National Research Agenda comprised of numerous steps and phases. The most important of these steps included the following.

Start-up phase
A detailed action plan appeared in the first half of March 2015, fulfilling one of the mandate requirements. The action plan was amended a number of times in the course of the development process in the light of cumulative insights or in response to altered schedules and principles. In the end, an organic approach was adopted approach developed organically and many of the activities and initiatives came about spontaneously, responding to evolving circumstances and opportunities. A virtual environment (base camp) situated in the secretariat provided for the necessary convergence, sharing, and cooperation on projects.

The start-up phase included the construction of a website that functioned as a repository for all information concerning the Dutch National Research Agenda. The website was also used during the public consultations.

Public consultations
In keeping with the mandate, the Dutch National Research Agenda was not an exclusively institutional product, a decision taken primarily to clear the way for innovation. To respond as fully as possible to the Minister's wish that the Agenda should 'appeal to the imagination', and to generate maximum support for the Agenda, the Knowledge Coalition decided to embark on a broad public consultation procedure in which scientists, businesses, governmental and civil society organisations, and individual citizens could provide input.

Public consultations were rolled out in April with the help of a digital module. The public were invited to 'ask a scientist a question'. All residents of the Netherlands could submit questions on the website of the Dutch National Research Agenda, along with an explanation, a few key words, and their email address. No less than 11,700 questions were submitted.

Assessment and clustering of the questions
The initial intention was to assess the suitability of all submitted questions. This task was entrusted to the Royal Netherlands Academy of Arts and Sciences (KNAW) and The Young Academy as independent organisations with the requisite expertise. The Academy and The Young Academy appointed five juries for this purpose, analogous to the five broad areas of science that fall within the Academy's remit (Humanities, Life Sciences, Natural Sciences, Social Sciences, and Technical Sciences). The Steering Committee decided on the composition of the juries, which represented all organisations participating in the Knowledge Coalition.

With so many questions having been submitted, however, there was a change of plans. Instead of an assessment of each question, the questions were clustered and aggregated. The first step was to cluster the questions using intelligent software. The juries then assessed the resulting clusters and reorganised them into a set of 248 clusters. Each cluster was provided with an overarching main question and a brief explanation. In formulating these cluster questions, the juries adhered to the following guidelines:
1 research into the question had to be possible within a ten-year period;
2 the question had to be challenging and ground-breaking in nature; and
3 there had to be prominent Dutch research groups capable of examining the question, or conversely, convincing arguments for building such capacity.

Conferences
Three conferences were held in June. In keeping with the mandate for the Agenda, the conferences focused on 'science4science', 'science4competiveness', and 'science4society'. Their purpose was to bring further order to the 248 clusters, to add relevant information, and to further aggregate the questions where possible, based on these three perspectives.

A total of 900 persons attended the conferences. The attendees participated in disciplinary and multidisciplinary discussion groups in several different rounds. The outcomes of the conferences were documented in three reports that were submitted to the Steering Committee in early July.

Writing and editing process
During the summer period, the Academy's juries aggregated the cluster questions more extensively based on the outcomes of the conferences. The result was a set of 195 cluster questions. The Steering Committee's aim, however, was to have a National Research Agenda consisting of no more than 150 questions. The Steering Committee therefore appointed a writing group and an editorial panel charged with reducing the number of clusters and refining the questions. The editorial panel was made up of members of the Knowledge Coalition; the writing group consisted mainly of secretariat staff nominated by the members of the Knowledge Coalition.

The writing group proposed to further aggregate the 195 questions into 140 cluster questions, based on the conference outcomes and in consultation with the editorial panel. All questions were also recast into a fixed format, including an explanation of the question itself, a demonstration of the connective power of the question (establishing connections between different disciplines and sectors, between various types of research from basic to applied, and between various research aims), and examples of the diversity of underlying questions submitted by the public.

Connections with existing agendas
From March to September 2015, the secretariat compiled a survey of existing research and policy agendas pursued by research institutions, governmental and civil society organisations and linked these agendas to the Dutch National Research Agenda questions.

The survey was the result of desk research. The secretariat searched the organisations' websites for research themes and priorities. One problem encountered was that there were major differences between research descriptions in terms of level of aggregation. To do justice to the various organisations, the secretariat worked exclusively with the organisations' own texts. Links to source pages were also included in the list. As a next step, the 140 cluster questions of the Dutch National Research Agenda were linked to the organisations' priorities.

Routes through the Dutch National Research Agenda
Between July and October, the focus was on framing and editing the questions. The structure of the final result also gradually became clear. The authorities had expected the Dutch National Research Agenda to identify a small number of priority research themes for policymaking and funding purposes. However, it quickly became clear that identifying only a small number of themes would do no justice to the depth and diversity of questions, nor to the very broad scope of existing research. The idea of plotting routes through the Dutch National Research Agenda arose during the conferences as a way of exploiting the depth of the 140 questions and fulfilling the mandate to make connections.

A route is a collection of related cluster questions that focus on a complex social, scientific or economic issue. While cluster questions connect original questions, routes connect the 140 cluster questions and other research and policy agendas by linking the questions to these agendas. A route is an instrument that allows users to approach a subject from different perspectives and discover which research groups are already working on it or which governmental or civil society organisations regard it as important. Routes can also help in the search for multi-sector and multidisciplinary research partners. 16 example routes that offer opportunities to make new connections were plotted out and incorporated by the Steering Committee in the Dutch National Research Agenda.

The Dutch National Research Agenda, on paper and digital
Once it had been decided *what* the Knowledge Coalition would produce – i.e. 140 cluster questions and 16 example routes – the next important question was *which form* the Dutch National Research Agenda would take. The answer was both a paper and a digital version. The digital version was considered to have various advantages: it would be easy for the Dutch public to access, and it would simplify management and updating. A digital environment would also allow users to get the most out of the dynamic routes.

The paper version of the Dutch National Research Agenda consists of an introduction that explains its aim and structure, the 140 cluster questions, and the 16 example routes. The 140 cluster questions are divided into five chapters:
– Man, the environment, and the economy;
– Individual and society;

- Sickness and health;
- Technology and society;
- Fundamentals of existence.

It concludes with a number of appendices that report public consultation statistics, provide a list of research and policy agendas, and describe the relationship between the 140 cluster questions and ten themes borrowed from the EU's Horizon 2020 programme.

The paper version of the Dutch National Research Agenda was presented to the authorities in November 2015. The digital version went live at the same time. At that point, it consisted of the original questions linked to the 140 cluster questions, which in turn were connected to the survey of existing research and policy agendas. It also consisted of the 16 example routes. One of the aims of the follow-up (see below) is to refine and extend the digital version of the Dutch National Research Agenda and to promote its usage for various purposes.

Special communication activities
Since early summer 2015, numerous special communication activities have been undertaken to raise familiarity with the Dutch National Research Agenda amongst the general public. This has promoted exchange between society and the research landscape and enhanced public support for science at large and the research agenda in particular.

'In Conversation'
Starting in early July, the possibility was created for the secretariat of the Dutch National Research Agenda to put organisations in touch with persons who had submitted a question concerning a theme relevant to the organisation's own field of activity. These organisations could then invite such persons to meetings, for example, or alert them to news of relevance to the subject of their question. For this purpose, the secretariat developed a digital tool that allowed organisations to approach persons who had submitted questions without violating their privacy. The tool gave those who had submitted questions the opportunity to communicate with researchers and other parties who shared their interests.

By the time the Dutch National Research Agenda was released, more than half of those who had submitted questions had received invitations to lectures, public meetings, and online forums of all kinds from a range of different organisations. Participating organisations included the National Institute for Public Health and the Environment (RIVM), the Royal

Netherlands Meteorological Institute (KNMI), Utrecht University, and the Royal Holland Society of Sciences and Humanities (KHMW).

Lowlands Science
Lowlands Science was an alliance between the Lowlands music festival organisation, Campagnebureau BKB, *New Scientist* magazine, the Royal Academy, and the Dutch National Research Agenda organisation. Its aim was to make science comprehensible for the general public, and it was organised during the Lowlands festival (21 to 23 August 2015). Several months prior to the event, an invitation to submit research proposals was distributed among scientists, universities, and research groups. The best proposals were presented daily at Lowlands. The NWA organisation invited a number of persons who had submitted questions to attend Lowlands Science and to put their questions to the researchers present that day. The invitation received a huge response. The secretariat filmed two encounters between individuals and scientists. They can be found at www.wetenschapsagenda.nl.

Living Room Lectures
The secretariat of the Dutch National Research Agenda cooperated with the Ministry of Education, Culture and Science on organising seven 'living room lectures' during the National Science Weekend. The living room lectures focused on submitted questions that had already been answered. Those who had posed questions welcomed a scientist into their home to discuss and answer the question, sometimes in the presence of a small audience. The Science Minister and State Secretary attended two of the lectures. Interested viewers could watch a live stream of the living room lectures in Periscope.

Society of Arts
Filmmaker Inge Meijer was commissioned by the Society of Arts and the Dutch National Research Agenda organisation to produce a film about the Agenda highlighting the role of those who had submitted questions. Meijer's aim was to film meetings between such individuals and scientists to show, at a micro level, the essence of the Dutch National Research Agenda: the convergence of science and society. Her film featured a number of living room lectures. It premiered on 29 November 2015 during the EUREKA! Festival in Amsterdam.

Besides Meijer's film, two other filmmakers produced films inspired by the questions submitted. Dutch poet laureate Anne Vegter composed a poem about the Dutch National Research Agenda. Finally, artist Koert van Mensvoort produced 'in vitro ice cream' that was served at the EUREKA! Festival to get participants thinking about food and sustainability.

EUREKA! Festival

The EUREKA! Festival, held on Sunday 29 November 2015, showcased science in all its many facets. Following the official presentation, this event unveiled the Dutch National Research Agenda for the general public.

The festival was held in Amsterdam and attracted about 3000 visitors. The festival programme was the result of collaboration with the communication departments of the various Knowledge Coalition partners and their organisations. A number of research universities adopted parts of the programme, the universities of applied science made a substantial contribution with their 'innovation catwalk', and The Young Academy filled one of the festival locations. The Society of Arts chose the festival to premiere its film about the Dutch National Research Agenda and delegated artists to reflect on the questions that had been submitted. Nijmegen's 'InScience' film festival organisation scheduled the remaining programme of science films. Shell, Unilever, and other businesses also cooperated.

Books

Publisher Nijgh & Van Ditmar published a book by science journalist Malou van Hintum on the development of the Dutch National Research Agenda, entitled *Wat wil Nederland weten?* (What does the Netherlands want to know?). Kennislink published a book answering a number of the questions posed.

Follow-up

The Dutch National Research Agenda – and especially the digital version – helps individuals and organisations find research partners that will enhance their efforts. To further this process, a series of 'route workshops' have been scheduled (starting in late 2015 and continuing in 2016) during which potential partners can explore the possibility of plotting new routes or elaborating on existing ones. To align the routes with existing agendas as closely as possible, the organisations will be asked to refine and maintain the connection between the Dutch National Research Agenda and the existing agendas.

One of the aims of the route workshops is to continue prioritizing the themes within the Dutch National Research Agenda. The Knowledge Coalition will use the outputs of these workshops as input for a manifesto advocating an integrated science, technology, and innovation policy that it will submit to the Dutch government.

The Knowledge Coalition does not regard the present 'product' as the finish line, but rather as the start of a revitalized and enhanced partnership, not only between its own members but also between other parties in Dutch society that have a deep interest in research. It is important to update the Dutch National Research Agenda at regular intervals in order to continue pursuing the current strategy, anticipate new developments, and above all maintain the momentum and support that has been generated for the current Agenda.

Reflections in retrospect

The process of formulating the Dutch National Research Agenda was once described as 'building an aircraft while in full flight'. The scale of the mandate and its expressed level of ambition, the composition of the Knowledge Coalition, the limited time available (nine months), and the innovative nature of the procedure made the process complicated and stressful. A number of underlying principles also raised the bar for those in charge: *everyone* was invited to contribute their input; none of those who had submitted questions should come away disappointed.

The chairpersons, the Steering Committee, and the secretariat have worked hard on the process and are satisfied with the result: a Dutch National Research Agenda consisting of 140 questions, 16 'exemplary routes', unexpected cross-connections, and a great deal of publicity for science – and especially Dutch science.

The mandate
The Knowledge Coalition's mandate was multifaceted in nature. The Dutch National Research Agenda was set up to encourage cooperation, unexpected connections, and imagination; to align research more closely to social and economic opportunities and requirements; to reflect and influence existing agendas; to demonstrate the excellence of Dutch research, make breakthroughs possible, and in doing so boost the international position of Dutch research; to have the support of the general public; and to make choices. Looking back on the process, the Steering Committee feels that it has successfully fulfilled this mandate.

Governance
Cooperation within the Knowledge Coalition – and the advice of the Liaison Group representing various segments of society at large – turned out to be an important prerequisite for the process leading to the Dutch National Research Agenda. All stakeholders were involved. They came to understand each other better, and to acknowledge their shared interests, including in the longer term. The members of the Knowledge Coalition did however have highly diverse organisations behind them that wanted to be involved and acknowledged. This made decision-making difficult at certain points.

As an independent party, the chairpersons were free to appear in the media and in other external contexts. Their activities generated support for the Dutch National Research Agenda and brought it into the limelight. The secretariat – consisting of representatives of the Knowledge Coalition – fast-tracked cooperation within the coalition; the representatives benefitted from each other's expertise and their shared aim stood above those of the separate parties.

Both the Steering Committee and the secretariat maintained innovative working methods. Although they initially adhered to the action plan, the process of decision-making and follow-up gradually became more organic. This approach created scope for creativity and unanticipated inspiration that enriched the outcomes of the process. The secretariat's method – working on projects in a virtual environment and making use of each other's complementary expertise – made it possible to facilitate and anticipate the cumulative insights of the Steering Committee and chairpersons.

Public consultations, assessment, and conferences
The public consultation procedure got many Dutch people from outside the scientific community involved in the Dutch National Research Agenda. The number of questions submitted exceeded expectations. The Steering Committee came to realise that it is rather difficult to manage processes closely during a public consultation procedure. No one could say how useful the outcomes of public consultation would be. Those who submitted questions did not know what would be done with their input, and scientists feared that ordinary citizens would decide what research they would be undertaking.

The Academy and The Young Academy made a valuable contribution by managing the task of assessment. In part thanks to their authority, their deep roots in science and innovation, and the meticulous way in which they clustered and aggregated the questions submitted, they ensured that the cluster questions would be framed in properly scientific terms, and that those who submitted questions would recognize their input in the relevant clusters.

The conferences proved to be excellent occasions for bringing together scientists, businesses, and society to discuss the juries' output. The attendees – approximately 900 in all – helped aggregate the cluster questions more precisely and contributed to their interdisciplinary nature.

Communication
The communication activities concerning the Dutch National Research Agenda sparked a huge response, as was evident from the almost 12,000 questions that were submitted. The activities also led to interaction between different disciplines, businesses, and civil society organisations, and between scientists and the general public. The various parties engaged with one another at different meetings and forums. Their interaction is a valuable outcome of this process.

Time frame
The timeline for this ambitious mandate was nine months. This did not deter the Steering Committee from launching various ambitious initiatives, such as the public consultation procedure and three major conferences. This was the first time ever that a national research agenda had been developed in this fashion, and it was uncharted territory for all those involved. That led to enormous creativity, but also put enormous pressure on the chairpersons, the Steering Committee, and the secretariat in every stage of the process. In addition, the broad spectrum of organisations represented in the Knowledge Coalition made rapid decision-making difficult. In the end, however, the mandate was fulfilled within the prescribed nine months.

Choices
As the process unfolded, it became clear to those involved that a national research agenda should in fact represent the full breadth of science. The task of choosing specific focus areas was decided against, first of all by the juries, who had no choice but to group as many questions as possible into valid clusters rather than select a specified number from among those submitted. This process continued during the conferences. It was unrealistic to expect sophisticated, well-argued choices fully supported by the participating organisations within the time remaining. With no prospect of additional funding, an important additional incentive for making choices was lacking.

The Steering Committee embraced the idea of the routes, meant to represent the connections between questions and parties. It will take more time to refine this idea, which will in fact lead to choices being made by means of a bottom-up process. After the release of the Dutch National Research Agenda, a series of route workshops will be organised whose outcomes will be presented to the relevant ministers and state secretary in mid-2016.

Innovation
The mandate given to the Knowledge Coalition had innovation as one of its key aims. Innovation was therefore a priority in the process. The Knowledge Coalition took an innovative approach in formulating 140 research questions illustrating the broad landscape of Dutch science and enjoying the support of the entire Knowledge Coalition. The process of enquiry also led to new products. The first is the digital version of the Dutch National Research Agenda, which shows the connections in that landscape, invites parties to enter into new alliances, and is constructed in a way that makes updating easy. The second is the idea of the routes, mentioned previously, which will serve as an impetus for bottom-up connections in the science and innovation sector. The third and final product is *In Conversation*, which will be continued after the completion of the Dutch National Research Agenda. *In Conversation* also illustrates the interaction that this process has generated between science and the public.

In addition to the Dutch National Research Agenda itself, the most important outcomes of the past nine months are the innovative process and the subsidiary products of that process.

Index

Academic scholarship 19
Agenda-led research 33-34, 36
Answers 9, 12, 20, 28, 42, 63, 70, 169, 176, 195, 197, 222
Applied research 34-36, 63, 70, 83-84, 90, 103-104, 106, 109-110, 113, 116, 130, 156, 168, 171, 173, 221, 223
Asking Questions 11-14, 16-17, 19-29, 32, 34-36, 40-43, 47, 51, 53-54, 57-58, 61, 64, 66-67, 70-72, 82, 84-85, 89, 101, 115, 127, 137-145, 147, 155, 163, 168-169, 172, 176, 186, 189, 193-194, 196-197, 200-203, 205-206, 209, 221-223, 228-238
Attitudes, ethics and habits 16, 175
Authority 37, 40-42, 98, 175, 178, 181, 214, 236

Basic research 33-34, 63-64, 70-71, 80, 86, 92, 101-104, 106-110, 113, 116-117, 137, 158, 221
Blue skies research 15, 34, 81
Bohr, Pasteur and Edison types of research 15, 104-105, 107-108, 108, 113, 116, 118

Central top-down steering mechanism 31
Citizens 9, 11, 14, 20-22, 27, 29, 32, 41, 54, 58, 115, 117, 133, 157, 168, 184, 189, 197-198, 205, 209, 221, 223, 229, 236
Citizens science *see Science*
Civil society organizations 9, 32, 96, 98, 124, 229-231, 237
Conflict of values 16, 181, 185
Curiosity 11, 17, 19-20, 22, 26-29, 33-36, 64, 82, 92, 98, 104, 110, 121, 127-128, 130, 199-200, 215, 221
Curiosity-driven research 33, 35, 127, 130, 200, 215

Disciplines 9, 15, 23-25, 27-28, 34-39, 43-44, 71, 76, 83-85, 88, 96, 110, 112, 127, 137, 155-156, 188, 199, 201-203, 209-210, 213, 221-222, 230, 237
Dutch National Research Agenda 9, 11, 13-14, 17, 20, 24-26, 28-29, 31-34, 36-37, 39-43, 47, 51, 56-58, 61-62, 67-68, 70-72, 76, 84, 89, 95, 101, 108, 115, 117-118, 124, 132, 144, 155, 157, 159, 161, 163, 165, 186, 209-215, 217, 221-223, 225-238

Economic value 35, 41, 131, 216

Free research 61, 66, 89, 111, 195
Funding 11, 26-27, 29, 32, 34, 36-37, 40, 42, 48, 55, 57, 65-67, 69, 72, 76-77, 80-81, 83, 86, 90, 93, 97-98, 102, 114-115, 119, 121-135, 140-141, 143-144, 147-153, 157, 159, 163, 165-167, 169, 174-175, 186, 188, 196, 202, 204, 210-212, 215-216, 222, 231, 237

Governmental organizations 32

Haldane Principle 13, 42, 85, 188
Humboldtina 'Bildung model 16, 181

Innovation 11-12, 15, 17, 31-33, 36, 39-40, 44, 47-59, 70, 75, 77-79, 97-98, 102-105, 107-108, 110-119, 121, 126-128, 131-132, 143, 145, 148-149, 159, 165-166, 182, 184-185, 193, 195, 209-211, 213-217, 221-224, 226, 229, 234-236, 238
  Innovation-oriented approaches 36
  Research and Innovation 11-12, 15, 43-44, 49-50, 55, 70, 96, 113, 115, 117-119, 182, 209-211, 214-217, 221-222, 224
Instrument 9, 13, 19, 27, 39-40, 47-51, 57, 64-65, 70, 94-96, 98, 112-113, 115, 123, 127, 131, 195, 198, 206, 213, 225, 231

Juries 23-25, 27, 32, 56, 229-230, 237

Knowledge Coalition 31-32, 39, 61, 132, 194-197, 206, 221-224, 226-227, 229-231, 234-238

Legitimacy 14, 16-17, 33, 40-41, 62, 67-68, 97, 129, 137-138, 142-144, 195
Lesson plan 28

Ministry of Education 24, 26, 29, 31, 44, 47, 65, 88, 91, 95, 99, 118, 122, 148, 156, 161, 166, 188, 225, 233

National Science Agenda *see* Dutch National Research Agenda
Netherlands Organization for Scientific Research 14, 75, 88, 95, 124, 130, 132, 141, 158, 194, 221, 227

Platform 11, 13-14, 27, 54, 115, 117, 119, 189, 202, 223, 228
Policymakers 14, 16, 34, 38, 42, 75, 78, 84-85, 106-107, 120, 124, 131-133, 140, 143-145, 151-152, 182, 202-203, 209, 211
Politicians 12, 16, 27, 29, 42, 67, 75, 78, 84, 88, 139, 159, 161, 188, 201, 214, 224
Practice-oriented research 14, 33-34, 61-68, 70-71
Priority-setting, communal forms of 193, 206
Processes 11, 49, 54, 67, 75, 78, 85, 98, 103, 112-113, 115, 118, 161, 194, 199-200, 210, 218, 236
  Bottom up 11, 13, 17, 21, 25-27, 31-32, 40, 43, 51, 56-57, 85, 111, 193, 209-212, 238
  Top Down 11
  Workshops 25, 222, 234-235, 238
Programming science *see* Science, programming

Public 11-15, 19-20, 28, 32, 37, 40-43, 47, 49, 55-59, 75, 77, 82, 84, 86, 114-115, 118-119, 127, 131, 133-134, 137-138, 142-145, 147, 151-152, 156-159, 164, 166, 168, 170, 177, 184, 188-190, 193-194, 196, 204-207, 209-212, 214-218, 228-238
Public domain 12-13
Public service 13, 187
Public support 14, 32-33, 40-43, 58, 232
Public voice 13

Questions 9-13, 16-17, 19-29, 32, 34-36, 40-43, 47, 51, 53-54, 57-58, 61, 64, 66-67, 70, 72, 82, 84-85, 89, 101, 115, 127, 137-145, 147, 155, 163, 168-169, 172, 176, 186, 189, 193-194, 196-197, 200-203, 205-206, 209, 221-223, 228-238

Rathenau Institute 15
Research
    Research budget 12, 16, 26, 89, 152, 156, 165
    Research institutes 12, 14, 31-32, 42, 50, 56, 85, 124, 157-158, 221
    Research and Innovation *see* Innovation, Research and Innovation
Researchers *see* Sience, scientists
Routes 14, 25, 27, 40, 51, 53-54, 57, 67, 84-85, 101, 117-118, 141-142, 155, 184, 194, 202, 222, 231-232, 234-238

Scepticism 16, 32-33
Science
    Citizen science 11, 13-14, 27, 115
    Collaborative science 25
    Science for competitiveness 24, 32, 82
    Science for society 24, 32, 82
    Science policy 12-13, 15-16, 31, 42-43, 47-48, 56-57, 75-76, 82, 84-85, 87-91, 95-98, 124, 133, 141, 156, 161, 165-166, 183, 193

Science, programming 13, 40, 61, 69-70, 97, 101, 111, 113, 210-212
Scientists 9, 14, 16, 19-20, 22-27, 29, 31-32, 41, 43, 63, 75-76, 79, 81, 83-85, 88-92, 97-98, 103, 115, 128, 130, 132-133, 137-145, 147-153, 157, 159, 161, 166, 169-170, 172-177, 182, 187-188, 193, 205, 229, 233, 236-237
Steering science 13-14
Societal relevance 35, 41, 130-131
Society 9, 12-17, 21, 24-26, 28-29, 31-32, 34, 38, 40-41, 43-44, 52-54, 62, 64, 67, 70-71, 75, 77-78, 82, 85, 87, 89-90, 94, 96-98, 101, 110, 112, 119, 123-124, 130-133, 137-140, 142-144, 155, 159, 168, 174-175, 178, 181-182, 184-187, 189, 194, 196-198, 200-201, 203, 206, 209, 211, 215-216, 221, 224-226, 229-237
Stakeholders 15, 27, 31, 77, 82-83, 85, 97, 116-117, 132, 134, 138-139, 141-142, 144, 194, 196-197, 201, 214, 216-217, 236
Strategic research 51, 56, 147
Systems 53, 59, 66, 71, 76-77, 80, 82, 86, 102-103, 114, 144, 149, 158, 198

Universities 12, 14-15, 31-32, 41, 50, 55-57, 61-72, 77, 80, 85, 88, 93-94, 101, 105, 109-110, 112-116, 118, 122-124, 128-129, 131, 148, 156-158, 160-163, 165, 174, 183-186, 188-189, 193-195, 202-207, 210, 218, 221, 223, 233-234
Universities of applied science 9, 61-72, 77
Untied research 35-36, 42, 187, 210
Utilitarian 'goose with the golden eggs' model 16

Valorisation 34, 88, 131, 158